Wildlife Conservation - Natural and Human-Influenced Ecosystems

Edited by Levente Hufnagel

Published in London, United Kingdom

Wildlife Conservation - Natural and Human-Influenced Ecosystems
http://dx.doi.org/10.5772/intechopen.1001786
Edited by Levente Hufnagel

Contributors
Beevans B. Biseko, Carlie Stransky Ogilvie, David Kaberege, Elhady A.M. Omer, Felician E. Chemihanda, Ferenc Mics, Frank F. Mallory, Graeme A. Spiers, Haythem Hamden, Kamel Charaabi, Levente Hufnagel, Lubna M.A. Hassan, Mariam Boudabouz, Meriem Msaad Guerfali, Michael G. Gardner, Mohamed Chaouachi, Musa M. Bulim, Philipo J. Mtweve, Salma Fadhl, Sana Bouzembila

First published in London, United Kingdom, 2025 by IntechOpen
IntechOpen is the global imprint of INTECHOPEN LIMITED, registered in England and Wales, registration number: 11086078, 167-169 Great Portland Street, London, W1W 5PF, United Kingdom

For EU product safety concerns: IN TECH d.o.o., Prolaz Marije Krucifikse Kozulić 3, 51000 Rijeka, Croatia, info@intechopen.com or visit our website at intechopen.com.

British Library Cataloguing-in-Publication Data
A catalogue record for this book is available from the British Library

Wildlife Conservation - Natural and Human-Influenced Ecosystems
Edited by Levente Hufnagel
p. cm.

This title is part of the Environmental Sciences Book Series, Volume 27
Topic: Environmental Resilience and Management
Series Editor: J. Kevin Summers
Topic Editor: Jose Navarro-Pedreño

Print ISBN 978-1-83769-802-8
Online ISBN 978-1-83769-801-1
eBook (PDF) ISBN 978-1-83769-803-5
ISSN 2754-6713

IntechOpen Book Series
Environmental Sciences
Volume 27

Aims and Scope of the Series

Scientists have long researched to understand the environment and man's place in it. The search for this knowledge grows in importance as rapid increases in population and economic development intensify humans' stresses on ecosystems. Fortunately, rapid increases in multiple scientific areas are advancing our understanding of environmental sciences. Breakthroughs in computing, molecular biology, ecology, and sustainability science are enhancing our ability to utilize environmental sciences to address real-world problems.

The four topics of this book series - Pollution; Environmental Resilience and Management; Ecosystems and Biodiversity; and Water Science - will address important areas of advancement in the environmental sciences. They will represent an excellent initial grouping of published works on these critical topics.

Meet the Series Editor

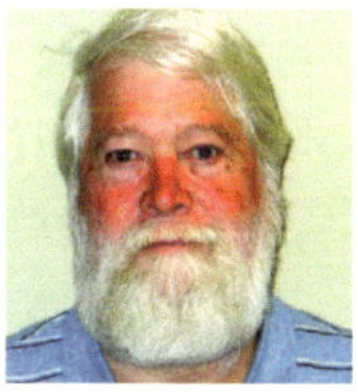

J. Kevin Summers is a Senior Research Ecologist at the Environmental Protection Agency's (EPA) Gulf Ecosystem Measurement and Modeling Division. He is currently working with colleagues in the Sustainable and Healthy Communities Program to develop an index of community resilience to natural hazards, an index of human well-being that can be linked to changes in the ecosystem, social and economic services, and a community sustainability tool for communities with populations under 40,000. He leads research efforts for indicator and indices development. Dr. Summers is a systems ecologist and began his career at the EPA in 1989 and has worked in various programs and capacities. This includes leading the National Coastal Assessment in collaboration with the Office of Water which culminated in the award-winning National Coastal Condition Report series (four volumes between 2001 and 2012), and which integrates water quality, sediment quality, habitat, and biological data to assess the ecosystem condition of the United States estuaries. He was acting National Program Director for Ecology for the EPA between 2004 and 2006. He has authored approximately 150 peer-reviewed journal articles, book chapters, and reports and has received many awards for technical accomplishments from the EPA and from outside of the agency. Dr. Summers holds a BA in Zoology and Psychology, an MA in Ecology, and Ph.D. in Systems Ecology/Biology.

Meet the Volume Editor

Dr. Levente Hufnagel is a full professor and the head of the Research Institute of Multidisciplinary Ecotheology of the John Wesley Theological College (Budapest, Hungary), working on ecology, biogeography, ecological research methodology, ecotheology and sustainability with over 20 years of experience at leading Hungarian academic institutions teaching Ph.D., MSc and BSc students from various social and cultural backgrounds, having more than 260 scientific publications (in both aquatic and terrestrial ecological aspects of plants, animals and microbes, at the community as well as population level) and more than 1500 independent citations. As a participant in several big ecological research and development projects, he has significant experience in multidisciplinary cooperation (with more than 200 coauthors in different publications). As the supervisor of several Ph.D., BSc and MSc theses and as editor-in-chief of an international scientific journal indexed by Web of Science and Scopus, he has wide supervising and editing experience. Dr. Hufnagel graduated from Eötvös Lorand University in Hungary with a Master's degree in Ecology and Evolutionary Biology and Ph.D. degree in Hydrobiology; he also graduated from Szent István University in Hungary with Ph.D. degree in Agricultural Science and with other degrees from the Corvinus University of Budapest and Adventist Theological College.

Contents

Preface

The explosive growth of the global population of humanity is rapidly reducing the extent of natural and semi-natural habitats and, thus, the chances of survival of the living world living there. Our planet's biodiversity is our common treasure, the preservation of which is not only an ethical issue and our moral duty but also our economic and scientific interests. Moreover, the rapid decline in biodiversity and the wave of mass extinctions are starting ecological processes that threaten the survival of civilizations. However, preserving biodiversity is no longer possible in protected natural areas alone, it requires us to use agricultural areas and urban ecosystems sustainably and manage them, taking into account nature conservation aspects so that they can also provide a refuge for endangered species.

This volume presents case studies that illustrate current problems, challenges, and difficulties in this area.

I hope that sharing experiences through such case studies will help both professional and amateur colleagues, researchers, government officials and activists committed to nature protection.

I want to express my sincere gratitude to the IntechOpen staff, Iva Horvat and Filip Lovricevic, for their high-quality professional work, help and support.

Levente Hufnagel
Research Institute of Multidisciplinary Ecotheology,
John Wesley Theological College,
Budapest, Hungary

Chapter 1

Introductory Chapter: Challenges and Tasks of Wildlife Conservation in Natural and Human-Influenced Ecosystems

Levente Hufnagel and Ferenc Mics

1. Introduction

The global human population has experienced a remarkable surge in numbers over the last two centuries, reaching a staggering figure of over eight billion individuals by November 2022 [1]. Consequently, this rapid expansion and urbanization pose a significant threat to the habitats and survival of various wildlife species worldwide, particularly animals and plants that face displacement due to land development or exploitation for human needs, such as food production. Additionally, the introduction of invasive species from different regions, climate change, pollution, hunting, fishing, and poaching further exacerbate the challenges faced by wildlife [2].

The preservation of wildlife offers a multitude of advantages to the ecosystem, encompassing the preservation of biodiversity, the regulation of ecological processes, the enhancement of ecosystem health, and the provision of economic benefits [3]. The International Union for Conservation of Nature and Natural Resources (IUCN) has identified over 25,000 species across the globe that are currently at risk and listed on the Red List of threatened species [4]. Various initiatives are currently being implemented to safeguard wildlife and the environment, such as the establishment of protected areas and national parks, the implementation of wildlife management and reintroduction schemes, the promotion of education and public awareness initiatives, as well as the enforcement of laws and regulations [5].

Previous research has demonstrated that the manner in which information is presented can elicit diverse reactions among the audience [6]. In order to effectively safeguard wildlife, it is crucial to comprehend the intricate dynamics of species interactions within their ecosystems, as well as the impact of environmental and human factors on their well-being [7].

In addition to passive and active, *in situ* and *ex situ* species protection, as well as passive in situ habitat protection, which previously prevailed in nature conservation, active in situ habitat protection (habitat management) and, due to climate change, active *ex situ* habitat protection (eco-engineering, habitat resettlement, and creation of ecological corridors), to extended biodiversity protection efforts (establishment of buffer zones and use of non-protected areas for maintaining biodiversity) are becoming more and more important. Protection activities must also be extended to artificial ecosystems in the immediate vicinity of national parks and protected areas, by establishing buffer zones around protected areas, where agricultural activity should only be allowed in the form of chemical-free organic farming, the spread of urbanization must be stopped, and the establishment of mines and industrial areas must be limited, and the protection of biodiversity must also be ensured in settlement areas (by establishing parks, artificial lakes, green roofs, tree rows, lawn areas, and methods of active species protection) [8].

2. Wildlife conservation in natural habitat

A variety of habitats exist to sustain a wide range of wildlife populations, encompassing forests, grasslands, wetlands, deserts, and marine ecosystems. The preservation of these habitats is advantageous not only for the species reliant on them but also for the general well-being of our Earth [9]. The presence of abundant biodiversity enhances the resilience of ecosystems, allowing them to effectively cope with and bounce back from various environmental stresses and disruptions [10]. Wildlife habitats face numerous threats that have a substantial impact on biodiversity and ecosystem health. It is crucial to comprehend these challenges in order to enact successful conservation strategies [11].

When habitats diminish in size and become fragmented, wildlife species encounter heightened challenges, including scarcity of resources, competition for survival, and decreased genetic variation. Consequently, this situation can result in declines in population numbers and, in more severe instances, the extinction of species [12]. Wildlife conservation relies heavily on the preservation and restoration of habitats as a fundamental strategy. Moreover, habitat restoration projects are essential in rehabilitating degraded ecosystems, thereby providing vital habitats for wildlife to thrive. These combined efforts are necessary to ensure that species have a safe refuge to fulfill their ecological roles effectively [13]. The wildlife conservation organization is in immediate need of implementing a more efficient and strategic methodology to tackle the current societal perspectives on wildlife and adapt to the evolving landscape and ecological circumstances [14]. One crucial step is to transition from a framework that primarily caters to a limited range of wildlife interests to a social-ecological paradigm. This paradigm entails adopting an approach to wildlife conservation that not only considers the ecological aspects but also incorporates the interests and involvement of a wider public. By doing so, the institution can ensure a more inclusive and comprehensive approach to wildlife conservation, thereby mitigating the decline of species and habitats [15]. An effective strategy should be based on fundamental principles that cover the duties and obligations of various stakeholders within the wildlife preservation sector, including elected and appointed officials (trustees), conservation experts working in governmental roles (trust managers), non-governmental organizations, and the general public [16].

Preserving valuable habitats for biodiversity necessitates additional measures beyond the mere establishment of protected areas. These measures encompass the implementation of robust legislation and the development of programs aimed at mitigating the various threats posed to undisturbed habitats [17]. Furthermore, it is essential to implement initiatives that promote sustainable socioeconomic development in regions that are not considered high-value habitats. However, there are certain jurisdictions where the laws and policies designed to protect habitats have been experiencing a decline in their effectiveness rather than being strengthened [18].

3. Urban wildlife conservation

The speed at which the process of urbanization is taking place globally is increasing [19]; consequently, there is a noticeable increase in the curiosity and fascination toward the wildlife that coexists within city environments [20]. Urban ecosystems are becoming more prevalent in today's society, and they have the potential to fulfill significant ecological functions [21]. This could involve implementing measures to ensure the preservation and sustainability of the ecosystem as well as the species that inhabit it [22].

Furthermore, apart from their ecological and conservation roles, urban green spaces offer significant advantages to human beings. Ecosystem services, such as pollination, are among the advantages that can be derived from this [23], air quality improvement [24], and carbon sequestration [25]. One of the most significant functions of urban ecosystems is their capacity to link city residents with their surroundings. As a result, the significance of urban conservation areas is expected to increase in the coming years [26]. Urban environments present distinct challenges due to their fragmented landscapes, that can isolate natural populations and hinder dispersal and functional connectivity. This fragmentation can result in demographic and genetic instabilities, as well as community-level impacts [27]. Organisms that inhabit urban areas can benefit from the establishment of corridors or stepping stone patches to facilitate their movement throughout cities [28]. The comprehensive incorporation of various life forms spanning different trophic levels, such as soil microorganisms, plants, invertebrates, and others, within the context of urban wildlife, has the potential to enhance our comprehension of urban ecosystem dynamics. This inclusive approach not only fosters impartial conservation efforts but also facilitates authentic human-nature interactions by embracing the diverse array of wildlife found in urban environments [29].

Wetlands, streams, and riparian areas frequently pose significant challenges for wildlife conservation efforts within urban and suburban landscapes. These aquatic ecosystems and their adjacent riparian zones are known for supporting a wide array of wildlife species, but they are also particularly vulnerable to the impacts of urban development [30, 31]. Nevertheless, certain species demonstrate resilience toward urbanization, and a few may even flourish in aquatic environments found in urban and suburban areas [32]. Effective management strategies for semiaquatic and aquatic wildlife in urban environments necessitate the implementation of appropriate stormwater treatment methods, habitat restoration initiatives, and potentially the reintroduction or translocation of species. Additionally, meticulous planning is crucial to safeguarding the preservation of vital habitats that still exist. By employing these strategies, not only will the welfare of aquatic and semiaquatic wildlife be safeguarded in urban and suburban regions, but the conservation of water quantity and quality will also yield benefits for human residents [33].

Author details

Levente Hufnagel* and Ferenc Mics
Research Institute of Multidisciplinary Ecotheology, John Wesley Theological College, Budapest, Hungary

*Address all correspondence to: leventehufnagel@gmail.com

References

[1] Pison G. World population: 8 billion today, how many tomorrow? Population & Societies. 2022;**604**(9):1-4

[2] Tilman D, Clark M, Williams DR, Kimmel K, Polasky S, Packer C. Future threats to biodiversity and pathways to their prevention. Nature. 2017;**546**:73-81

[3] Rands MRW, Adams WM, Bennun L, Butchart SHM, Clements A, Coomes D, et al. Biodiversity conservation: Challenges beyond 2010. Science. 2010;**329**(5997):1298-1303

[4] IUCN. The IUCN (International union for conservation of nature and natural resources) red list of threatened species, table 1: Numbers of threatened species by major groups of organisms. 2017. Available from: http://www.iucnredlist.org/about/summary-statistics#Tables_1_2

[5] Gavin MC, McCarter J, Berkes F, Mead ATP, Sterling EJ, Tang R, et al. Effective biodiversity conservation requires dynamic, pluralistic, partnership-based approaches. Sustainability. 2018;**10**(6):1846

[6] Ashby FG, Prinzmetal W, Ivry R, Maddox WT. A formal theory of feature binding in object perception. Psychological Review. 1996;**103**(1):165-192

[7] Willis KJ, Araújo MB, Bennett KD, Figueroa-Rangel B, Froyd CA, Myers N. How can a knowledge of the past help to conserve the future? Biodiversity conservation and the relevance of long-term ecological studies. Philosophical Transactions of the Royal Society B. 2007;**362**:175-186

[8] Hufnagel L. Biodiversity conservation levels and approaches in today's global ecological crisis. In: Hufnagel L, editor. New Insights into Protected Area Management and Conservation Biology. London, UK: IntechOpen; 2024. DOI: 10.5772/intechopen.113079

[9] Harte J. Land use, biodiversity, and ecosystem integrity: The challenge of preserving earth's life support system. Ecology Law Quarterly. 2011;**27**(4):A3

[10] Kovalenko KE, Thomaz SM, Warfe DM. Habitat complexity: Approaches and future directions. Hydrobiologia. 2012;**685**:1-17

[11] Bastille-Rousseau G, Wittemyer G. Characterizing the landscape of movement to identify critical wildlife habitat and corridors. Conservation Biology. 2020;**35**(1):346-359

[12] Yang H, Viña A, Tang Y, Zhang J, Wang F, Zhao Z, et al. Range-wide evaluation of wildlife habitat change: A demonstration using Giant Pandas. Biological Conservation. 2017;**213**:203-209

[13] Possingham HP, Bode M, Klein CJ. Optimal conservation outcomes require both restoration and protection. PLOS Biology. 2015;**13**(1):e1002052

[14] Fischman RL, Adamcik RS. Beyond trust species: The conservation potential of the national wildlife refuge system in the wake of climate change. Natural Resources Journal. 2011;**51**(1):1-33

[15] Decker D, Smith C, Forstchen A, Hare D, Pomeranz E, Doyle-Capitman C, et al. Governance principles for wildlife conservation in the 21st century. Conservation Letters. 2016;**9**(4):290-295

[16] Sagarin RD, Turnipseed M. The public trust doctrine: Where ecology meets natural resources management. Annual Review of Environment and Resources. 2012;**37**:473-496

[17] Abegão JLR. Where the wild things were is where humans are now: An overview. Human Ecology. 2019;**47**:669-679

[18] Appeldoorn RS, Lindeman KC. A Caribbean-wide survey of no-take marine reserves: Spatial coverage and attributes of effectiveness. Gulf and Caribbean Research. 2003;**14**(2):139-154

[19] Wang H, He Q, Liu X, Zhuang Y, Hong S. Global urbanization research from 1991 to 2009: A systematic research review. Landscape and Urban Planning. 2012;**104**(3-4):299-309

[20] Soulsbury CD, White PCL. Human–wildlife interactions in urban areas: A review of conflicts, benefits and opportunities. Wildlife Research. 2015;**42**(7):541-553

[21] Collins MK, Magle SB, Gallo T. Global trends in urban wildlife ecology and conservation. Biological Conservation. 2021;**261**:109236

[22] Hunold C, Mazuchowski M. Human–wildlife coexistence in urban wildlife management: Insights from nonlethal predator management and rodenticide bans. Animals. 2020;**10**(11):1983

[23] Davis AY, Lonsdorf EV, Shierk CR, Matteson KC, Taylor JR, Lovell ST, et al. Enhancing pollination supply in an urban ecosystem through landscape modifications. Landscape and Urban Planning. 2017;**162**:157-166

[24] Roeland S, Moretti M, Amorim JH, Branquinho C, Fares S, Morelli F, et al. Towards an integrative approach to evaluate the environmental ecosystem services provided by urban forest. Journal of Forestry Research. 2019;**30**:1981-1996

[25] Mexia T, Vieira J, Príncipe A, Anjos A, Silva P, Lopes N, et al. Ecosystem services: Urban parks under a magnifying glass. Environmental Research. 2018;**160**:469-478

[26] Hobbs SJ, White CL. Achieving positive social outcomes through participatory urban wildlife conservation projects. Wildlife Research. 2016;**42**(7):607-617

[27] Gomes V, Ribeiro R, Carretero MA. Effects of urban habitat fragmentation on common small mammals: Species versus communities. Biodiversity and Conservation. 2011;**20**:3577-3590

[28] Aziz HA, Rasidi MH. The role of green corridors for wildlife conservation in urban landscape: A literature review. IOP Conference Series: Earth and Environmental Science. 2014;**18**:012093

[29] Keniger LE, Gaston KJ, Irvine KN, Fuller RA. What are the benefits of interacting with nature? International Journal of Environmental Research and Public Health. 2013;**10**:913-935

[30] Croteau MC, Hogan N, Gibson JC, Lean D, Trudeau VL. Toxicological threats to amphibians and reptiles in urban environments. In: Mitchell JC, Brown REJ, Bartholomew B, editors. Urban Herpetology: Herpetological Conservation. Salt Lake City, Utah: Society for the Study of Amphibians and Reptiles; 2008

[31] Solomon KR, Carr JA, Du Preez LH, Giesy JP, Kendall RJ, Smith EE, et al. Effects of atrazine on fish, amphibians, and aquatic reptiles: A critical review. Critical Reviews in Toxicology. 2008;**38**(9):721-772

DOI: http://dx.doi.org/10.5772/intechopen.1006656

[32] Tawari-Fufeyin P, Ekaye SA. Fish species diversity as indicator of pollution in Ikpoba River, Benin City, Nigeria. Reviews in Fish Biology and Fisheries. 2007;**17**:21-30

[33] Price SJ, Snodgrass JW, Dorcas ME. Managing aquatic environments for wildlife in urban areas. In: McCleery RA, Moorman CE, Peterson MN, editors. Urban Wildlife Conservation: Theory and Practice. London: Springer; 2014

Chapter 2

Overview on Genetic Diversity and Population Structure of the Nubian Ibex in Sudan: Key Aspects for Management and Conservation Strategies

Lubna M.A. Hassan and Elhady A.M. Omer

Abstract

The Nubian ibex is important for conservation strategies, population health, evolutionary history, genetic adaptations, and management and breeding programs. Understanding these aspects is crucial for targeted conservation strategies, species survival, and maintaining genetic diversity. This chapter focuses on the genetic diversity and population structure of the Nubian ibex. We compare it to related species and use genotyping data to analyze heterozygosity, differentiation, and cluster analysis. Through our research, we emphasize the significance of comprehending genetic variations for the purpose of implementing effective conservation efforts. These insights play a crucial role in the preservation of this remarkable species in Sudan, ultimately raising awareness about the importance of protecting the Nubian ibex for future generations.

Keywords: Nubian ibex, conservation strategies, genetic adaptations, genetic diversity, effective conservation efforts

1. Introduction

1.1 General overview

The Nubian ibex, scientifically identified as *Capra nubiana*, is among the ten species that fall under the genus Capra. The remaining species *encompass C. caucasica, C. cylindricornis, C. falconeri, C. pyrenaica, C. ibex, C. sibirica, C. walie, C. aegagrus,* and the domestic goat species *(C. hircus)*. The Nubian ibex can be found in fragmented populations scattered throughout Egypt, Sudan, Eritrea, Jordan, Oman, Yemen, and Saudi Arabia [1]. As per the classification provided by the International Union for Conservation of Nature's (IUCN) Red List, the Nubian ibex is categorized as "Vulnerable" [1]. This particular species has developed impressive adaptations to thrive in hot desert environments characterized by high diurnal temperatures and limited access to food and water resources [2]. One of their survival strategies involves the consumption of desert plants that are abundant in secondary metabolites such as alkaloids, which indicates the presence of an efficient detoxification system [3]. Given its resilience in challenging habitats, the Nubian ibex holds potential as a model species for studying genetic adaptations that could be beneficial for other economically valuable livestock species.

Ecologically, Nubian ibex are important for their impact on plant communities in their habitat through browsing [4]. They also serve as a key food source for predators like the Arabian leopard, contributing to the balance of their ecosystem despite the challenges they face [5].

The aim of this chapter is to provide a comprehensive analysis and action plan for the preservation of the endangered Nubian ibex species in Sudan. Understanding their current status, conservation strategies, relevant laws, and raising awareness about the importance of protecting endangered species are crucial steps in developing a detailed conservation plan. To establish effective breeding and conservation strategies for the Nubian ibex, it is imperative to collect data on genetic diversity within Nubian ibex populations and compare it with other Capra species. Additionally, the identification of casein protein variants in different domestic goat breeds from Sudan and wild goat species like the Bezoar ibex, Nubian ibex, and Alpine ibex is essential for conservation initiatives.

2. Materials

In order to compile information for the chapter review on the Nubian ibex in Sudan, an extensive literature review was conducted. The search process involved utilizing reputable academic databases such as Google Scholar and Web of Science. The methodology employed can be outlined as follows:

2.1 Literature search and screening

2.1.1 Data sources: Scholarly articles, reports, and studies pertaining to the Nubian ibex were accessed

2.1.2 Keywords

Various keywords were combined to enhance the search process, including terms such as "Nubian ibex," "Conservation strategies," "Genetic adaptations," "Genetic diversity," and "Effective conservation efforts."

2.1.3 Screening process

Articles were screened based on their relevance to the objectives of the study.

2.2 Grouping of reviewed papers

2.2.1 Objective-based arrangement

Reviewed papers were categorized according to the specific objectives of the study in Section 3 below.

2.2.2 Framework formation

This categorization served as the foundational framework for the subsequent analysis.

2.2.3 Sections 4, 5, and 6 below

These sections present case studies that further explore the research findings:

2.2.3.1 Genetic diversity of Nubian ibex

An examination of the genetic diversity of Nubian ibex in comparison to other ibex species and domesticated goats.

2.2.3.2 Capture sequencing for rare casein variants

The investigation of rare casein variants in Nubian ibex milk through capture sequencing.

2.2.3.3 Current conservation efforts in Sudan

An evaluation of the ongoing conservation initiatives aimed at protecting the Nubian ibex population in Sudan.

3. Methods for gathering and choosing sources

Methods for gathering and choosing sources for the review of our Nubian ibex chapter:

3.1 Source collection

3.1.1 Data sources

The study utilized respected academic databases like Google Scholar and Web of Science.

3.1.2 Types of sources

Scholarly articles, reports, and studies specifically focused on the Nubian ibex were consulted.

3.2 Source selection criteria

3.2.1 Relevance

Articles were evaluated based on their relevance to the objectives of the study.

3.2.2 Quality

Emphasis was placed on selecting high-quality, peer-reviewed sources.

3.2.3 Recentness

Priority was given to recent publications.

3.2.4 Coverage

A wide range of topics concerning Nubian ibex conservation, genetics, and adaptations were taken into account.

4. Background and context

4.1 Nubian ibex habitat

Nubian ibex are well-suited to survive in challenging environments such as canyons, rocky cliffs, and scree fields with loose rocks. They are adaptable to arid regions with limited vegetation, thriving at elevations ranging from sea level to an impressive 3000 meters. These remarkable animals are typically found in remote, elevated, and steep cliffs, where they demonstrate diverse habitat utilization patterns due to noticeable seasonal changes. Ibex have a preference for rugged and steep landscapes, with their evolutionary adaptation allowing them to flourish in high, mountainous regions where cliffs play a crucial role in their survival. They are agile climbers, using their skills to evade predators, and exhibit seasonal movements to different elevations, favoring higher slopes year-round. During the summer, they migrate to the highest points, descending to lower elevations in spring to take advantage of fresh plant growth. Nubian ibex primarily inhabit rocky desert mountains with steep slopes and hills, utilizing plateaus, canyons, and wadis within their habitat. Their diet consists of a wide variety of herbaceous and woody plants, as highlighted by [1, 6].

4.2 Nubian ibex geographical distribution range

The *Capra nubiana*, commonly known as the Nubian ibex, was once widely distributed in the mountainous regions of northeastern Africa and the Middle East (**Figure 1**). However, its current distribution is primarily concentrated east of the Nile in Egypt, northeastern Sudan, northern Ethiopia, western Eritrea, western Jordan, and scattered locations in western and central Saudi Arabia, Yemen, and southern Oman. It is important to note that this species is no longer found in Lebanon and Syria. In Sudan, the Nubian ibex exists in small, isolated populations in various locations within the Red Sea Hills. According to reports, these ibexes were previously documented to inhabit three protected regions along the Red Sea shoreline, namely the Erkawit and Sinkat Sanctuaries, as well as the Tokar Game Reserve. Additionally, they were also found in the northernmost area of the region, near the border with Egypt. The exact number of remaining ibexes in Sudan is uncertain, but reports suggest that it is unlikely to exceed a few hundred in the Red Sea Hills [6–12]. Despite the availability of some data from the Wildlife Research Center (WRC) and Wildlife Conservation General Administration (WCGA), a significant amount of information remains unpublished.

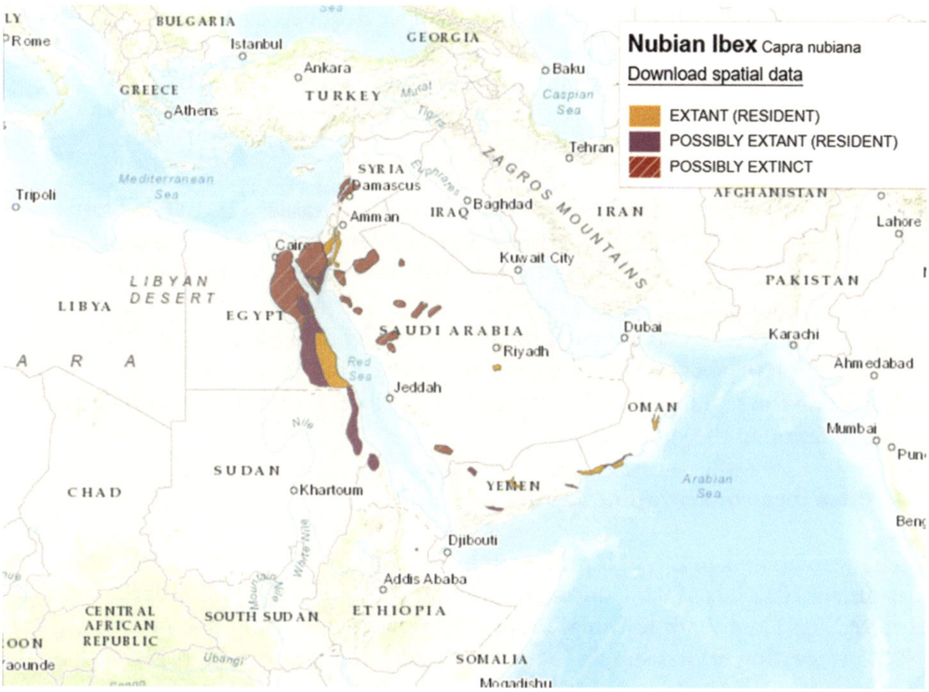

Figure 1.
IUCN (International Union for Conservation of Nature) 2020. Capra nubiana. *The IUCN red list of threatened species. Version 2023–1. Source: [1].*

4.3 Nubian ibex population

Population data for the Nubian ibex range is limited, especially in Sudan, Eritrea, Ethiopia, and Yemen. The global population is believed to be less than 5000 mature individuals [13, 14]. In Egypt, there are two main populations, with estimates of 400–1000 ibex in the Eastern Desert and 200–250 in South Sinai. Population trends in Egypt are declining. Unfortunately, there are no population estimates available for Eritrea or Sudan, and very few reports due to the recent unrest in the region. Hunting reports from before 2010 confirmed the scarcity of Nubian ibex north and south of Port Sudan [10]. It is challenging to determine the exact number of ibex remaining in Sudan, but reports suggest that it is unlikely to be more than a few hundred in the Red Sea Hills [10]. As for Eritrea and Ethiopia, there are no recent confirmed records, and the status of Nubian ibex in these areas remains unknown.

The Nubian ibex in Israel is found in the Judean Desert, Negev Desert Highlands, Eilat Mountains, and Golan Heights. The estimated population is around 800 in the Judean Desert and 400 in the Negev Desert Highlands [15]. In Jordan, a captive breeding program was started in 1989 to save the ibex population, and they can now be found in Wadi Mujib, Dana, and Wadi Rum [6, 16–19]. However, poaching is a major threat. In Oman, the ibex population is declining due to poaching, with no recent records in certain areas. The Dhofar region is a key habitat for ibex. The Dhofar region in Oman has the largest population of Nubian ibex, estimated at 600–1100 individuals. In Yemen, ibex are rare and facing human impacts, worsened by ongoing unrest [20]. In Saudi Arabia, ibex populations have declined outside protected areas, with a 75% decrease in one reserve since 2005 [21]. The trend for Nubian ibex populations is decreasing overall, with habitat and ecology facing challenges.

4.4 Nubian ibex conservation status

IUCN Status: The Nubian ibex is classified as "Vulnerable" [1]. CITES Status: Not listed under CITES [22]. Other Information: Further details are available in [6]. Country-Specific Regulations: Egypt: Protected under Agricultural Law 53/I 966, hunting prohibited. Israel: Fully protected, vulnerable in arid-zone habitat, rare in Golan Heights. Oman: Completely protected under Ministerial Decision No. 4 of 1976. The Nubian ibex is safeguarded by the United Arab Emirates' Federal Law No. 11 of 2002 on the regulation and protection of wildlife, prohibiting hunting, trading, or causing harm to the species.

4.5 Nubian ibex conservation actions

The Nubian ibex in Egypt is protected by Agricultural Law No. 53/l 966 and amendment 1012 July 1992. Hunting is strictly prohibited. The St. Katherine Protected Area and Wadi El Gimal Protected Area are important habitats for the species. Conservation efforts should focus on maintaining ibex corridors in South Sinai, increasing ranger patrols and enforcement in the Southern Red Sea Mountains, and establishing a camera trap monitoring program [14].

The current status of the Nubian ibex in Eritrea is unknown. Basic distributional data needs to be collected to understand the population. A survey is necessary to gather information about the ibex in Eritrea [23].

Similar to Eritrea, the status of the Nubian ibex in Ethiopia is unknown. Basic distributional data needs to be collected. A survey is necessary to understand the population.

The Israel Nature Reserve Authority has classified the ibex as "Vulnerable" in arid zones and "Rare" in Mediterranean regions like the Golan Heights. The ibex is fully protected in Israel, with over 80% of its habitat in protected areas. Ongoing management includes surveys and research to protect populations and habitats, focusing on population dynamics, genetics, habitat needs, and human disturbances. Some past suggestions to improve water resources have been implemented in certain areas [24–27].

The Nubian ibex has been protected since 1978, leading to population growth in the Wadi Mujib Nature Reserve. It can be found in three protected areas. Limited camera trap monitoring is ongoing [6, 16]. In Lebanon, the ibex has been eradicated, but a re-introduction program is underway. Ibex have been relocated from Jordan to the Chouf District, with plans for the Shouf Biosphere Reserve. In Syria, there are no conservation efforts, resulting in eradication. In Oman, the ibex is legally protected, but numbers are declining due to poaching. It can be found in the Jabal Samhan Nature Reserve and Al Wusta Wildlife Reserve.

The Nubian ibex population in Saudi Arabia is declining due to illegal hunting, despite a hunting bylaw in 1979. They are currently only found in the Ibex Reserve in Hawtat Bani Tamim. Efforts are being made to reintroduce them to other areas like the Sharaan Natural Reserve in Al Ula [21, 28]. Conservation efforts in Yemen are uncertain due to ongoing unrest. Collaborating with local communities is important to reduce hunting and ensure sustainability. Sustainable tourism projects in protected areas can also benefit local economies. Continuous population monitoring is crucial in Israel and Oman for ibex conservation.

Research is needed on the distribution of the Nubian ibex in countries like Sudan, Eritrea, Ethiopia, Yemen, Egypt, and southern Oman. Collaboration between these nations can improve data quality. Predictive modeling techniques can help address knowledge gaps and identify important conservation areas. Israel has extensively studied its ibex populations, focusing on metapopulation dynamics, environmental influences, survival rates, and human-induced mortality. Further investigation is needed to understand the behavioral effects of human presence and landscape fragmentation, as highlighted by [29].

4.6 Nubian ibex threats

Primary challenges faced by Nubian ibex populations include illegal hunting, limited enforcement against poachers, habitat degradation, competition with livestock, habitat loss and fragmentation, reduced connectivity between populations, Water scarcity and climate change further impact the ibex's access to water sources and overall survival. Disease outbreaks and climate change impacts like rising temperatures and altered precipitation patterns, and grazing livestock and land-use changes reduce vegetation cover and fragment the ibex's habitat, leading to reduced genetic diversity and limited movement. These threats pose a significant risk to the species' survival [1, 14, 15].

4.7 Why understanding its genetics and population structure is crucial

Understanding the genetics and population structure of the Nubian ibex (*Capra nubiana*) holds great significance for a variety of reasons.

Firstly, it is of utmost importance for conservation strategies. By gaining insights into the genetic diversity of the species, conservationists can better understand its ability to adapt to changing environments. This knowledge allows for the development of targeted strategies aimed at effectively preserving the Nubian ibex. In order to implement successful conservation efforts, it is crucial to possess a comprehensive understanding of the genetic composition of different populations.

Secondly, the genetics and population structure of the Nubian ibex play a significant role in ensuring population health. Genetic diversity is essential for the long-term survival of a species. Inbreeding and reduced genetic variation can lead to health issues and increase the species' vulnerability to diseases. By monitoring the population structure, conservationists can identify isolated or declining populations, enabling appropriate measures to be taken for their conservation.

Thirdly, comprehending the evolutionary history of the Nubian ibex is of utmost importance. Genetic studies provide valuable insights into the evolutionary relationships between the Nubian ibex and other related species. This knowledge is crucial in making informed decisions regarding conservation efforts.

Fourthly, genetic adaptations are key to the Nubian ibex's ability to thrive in arid and rocky habitats. Identifying specific genes that are related to adaptation is vital for conserving these unique traits that enable the species to survive in such challenging environments.

Lastly, genetic data plays a crucial role in management and breeding programs. It guides captive breeding programs, ensuring that genetic diversity is maintained within captive populations to prevent inbreeding and maintain the overall health and viability of the species.

5. Genetic diversity of Nubian ibex in comparison to other ibex and domesticated goat species

5.1 Genetic diversity assessment

5.1.1 The genotyping methods

5.1.1.1 SNP chip design

The GoatSNP50 chip, also known as the 52 K single nucleotide polymorphisms (SNP) chip, was developed for goats to create a moderate-density SNP assay suitable for various goat breeds. The success of this chip relies on reliable bioinformatic SNP detection, uniform distribution of SNPs across the genome, and appropriate selection of Minor Allele Frequencies (MAF) for diverse breeds [30].

5.1.1.2 SNP discovery and selection

The International Goat Genome Consortium brought together three SNP discovery projects, resulting in the identification of approximately twelve million high-quality SNP variants within and between six goat breeds. These SNPs were evenly distributed across the goat genome.

5.1.2 Manufacturing and validation

The selected SNPs were manufactured by Illumina, Inc. and published in db SNP, including flanking sequences and map positions on goat assemblies, sheep genome V2, and cattle UMD3.1 assembly. The chip was validated using ten different goat breeds, successfully genotyping 52,295 loci.

5.1.3 Efficient strategy

The efficient strategy involved whole-genome Next Generation Sequencing (NGS) for SNP discovery, mapping on a contig genome assembly for accurate genomic positions, and utilization of Illumina design tools for efficient chip production [30, 31].

5.2 Observed heterozygosity values for Nubian ibex, alpine ibex, and bezoar ibex

Heterozygosity values were observed for the Nubian, Alpine, and Bezoar ibex by analyzing available research data. The Nubian ibex exhibited a heterozygosity range of 0.02 to 0.44 with 33,698 SNPs and 662 SNPs. In comparison, the Alpine ibex displayed a range of 0.01 to 0.38 with the same number of SNPs. The Bezoar ibex showcased a heterozygosity range from 0.13 to 0.38 with both sets of SNPs. These results shed light on the genetic diversity within these ibex species, highlighting the Nubian ibex's relatively stable heterozygosity levels despite its endangered status [32].

5.3 Key findings from a study on Sudanese goat breeds, including Taggar goats, are

5.3.1 High genetic diversity

Using the 50 K SNP chip, researchers found that over 96% of the SNPs showed genetic variation within the population.

5.3.2 Healthy genetic diversity

Minimal evidence of inbreeding within each breed indicates a healthy level of genetic diversity.

5.3.3 Low variation between breeds

While there is diversity within each breed, the variation between different breeds is relatively low.

5.3.4 Unique genetic traits of Taggar goats

Genetic analysis revealed distinct genetic traits in Taggar goats that differentiate them from other breeds.

5.3.5 Identification of important genomic regions and candidate genes

The study identified specific genomic regions associated with growth, development, bone formation, and the immune system. Candidate genes related to these traits were also identified.

5.3.6 Implications for breed selection and improvement

These findings are valuable for informing breed selection programs and improving local breeds. The study's insights can contribute to the enhancement and preservation of Sudanese goat breeds (**Table 1**) [33].

5.4 Population differentiation

5.4.1 Pairwise FST values for differentiation between species

Comparisons between Bezoar ibex and Taggar goats yielded values ranging from 0.17 to 0.35. For Bezoar ibex versus Alpine ibex, the F_{ST} values fell between 0.47 and 0.91. When comparing Bezoar ibex with Nubian ibex, the F_{ST} values ranged from 0.33 to 0.90.

Population	*N*	H_O	H_E	F_{IS}	**PIC**
		33,698 | 662 SNPs	33,698 | 662 SNPs	662 | 33,698 SNPs	33,698 | 662 SNPs
Bezoar ibex	4	0.13 | 0.38	0.11 | 0.30	−0.2783 | −0.1171	28.1 | 77.5
Alpine ibex	8	0.01 | 0.38	0.01 | 0.22	−0.7165 | −0.5675	1.8 | 50.9
Nubian ibex	8	0.02 | 0.44	0.01 | 0.31	−0.3953 | −0.2063	4.3 | 81.2
Taggar	24	0.40 | 0.41	0.39 | 0.36	−0.1227 | −0.0264	99.5 | 99.8
Combined	44	0.23 | 0.41	0.39 | 0.43	0.0593 | 0.3998	

N number of examined individuals per species, H_O observed heterozygosity, H_E expected heterozygosity, F_{IS} Wright's inbreeding coefficient, PIC percentage of markers segregating within this population. Source: [32].

Table 1.
Estimated diversity metrics comparing species based on 33,698 SNPs and 662 SNPs.

5.4.2 Molecular variance analysis results

The molecular variance analysis indicated that approximately 74–78% of the variance could be attributed to differences between species.

5.5 Cluster analysis and principal component analysis

5.5.1 Describe the clusters detected based on Nei's genetic distance

Nubian and Alpine ibexes were grouped together, while Taggar goats and Bezoar ibexes formed another cluster.

5.5.2 Explain how principal component analysis refined these clusters

PCA analysis confirmed these groupings by showing genetic similarities and differences.

5.6 Novel SNPs and genomic regions

5.6.1 Highlight any novel single nucleotide polymorphisms (SNPs) discovered

Novel SNPs in the Nubian ibex and other species are continuously being discovered through ongoing research. These genetic variations play a crucial role in adaptation and evolution, particularly in arid and rocky habitats. High-throughput sequencing technologies aid in identifying these SNPs, which can be used as markers for various studies and initiatives. As genomic research advances, we can expect to uncover more novel SNPs that will deepen our understanding of genetic diversity and ecological interactions.

5.6.2 The role of milk protein genes in species differentiation

Milk protein genes help differentiate between species, indicating adaptations related to lactation, diet, and environment. Discovering these genes provides valuable information on domesticating wild Capra species. The Nubian ibex exhibits distinct genetic traits, with milk protein genes playing a pivotal role in species differentiation (**Figures 2–4; Table 2**) [32].

6. Capture sequencing to explore and map rare casein variants in Nubian ibex

This research focuses on the casein protein variants in goats, specifically the critically endangered Nubian ibex (*Capra nubiana*). The study utilized high-density capture sequencing to investigate and map rare casein variants across different goat breeds and wild goat species.

6.1 Background

Casein genes (CSN1S1, CSN2, CSN1S2, and CSN3) encode proteins that are present in milk. These proteins play a role in milk composition, cheese processing properties, digestibility, and human nutrition. Milk protein variants are crucial for breed characterization, biodiversity studies, and phylogenetic analyses.

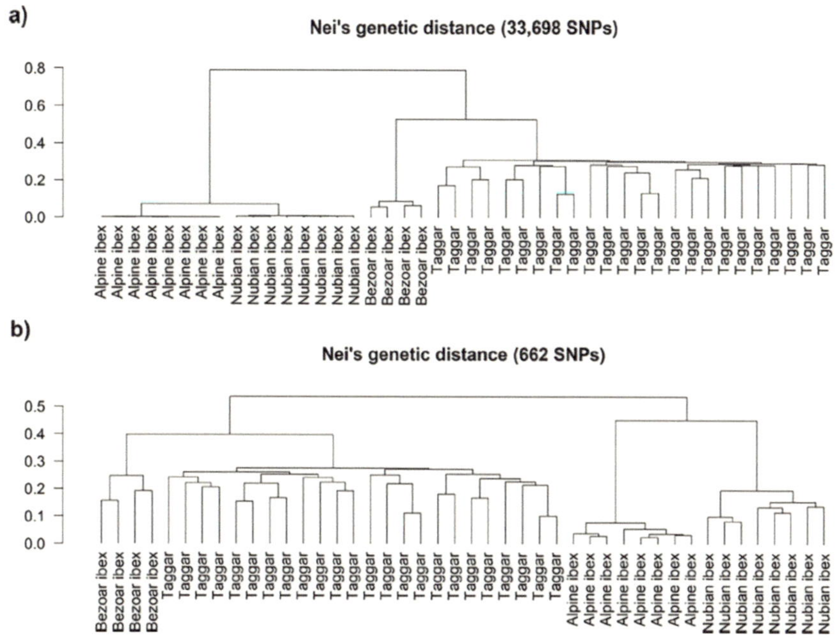

Figure 2.
Hierarchical clustering dendrogram of the individuals based on Nei's genetic distance. Hierarchical clustering based on 33,698 SNPs. Hierarchical clustering based on 662 SNPs. The Y axis represents the distance between individuals or clusters of individuals. The X axis represents every individual of the four species and shows clusters. Source: [32].

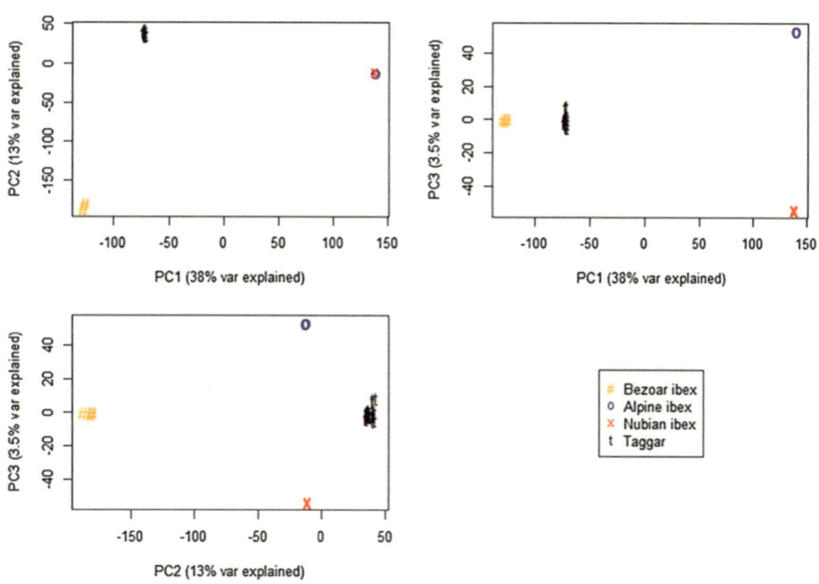

Figure 3.
PCA plot of the four goat species. Top left: scatter plot using the first and second principal components (PC1 vs. PC2). Top right: scatter plot using the first and third principal components (PC1 vs. PC3). Bottom left: scatter plot using the second and third principal components (PC2 vs. PC3). Source: [32].

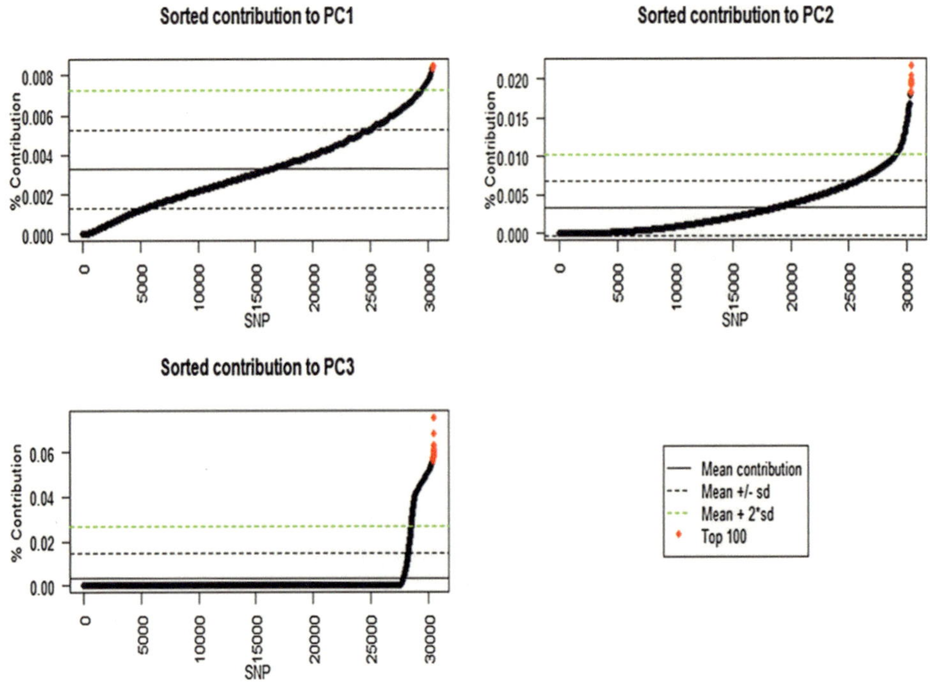

Figure 4.
SNP contributions to PCs sorted in ascending order. Top left: SNPs are sorted according to their contribution to the principal component in ascending order. Top left: SNPs contributing to PC1. Top right: SNPs contributing to PC2. Bottom left: SNPs contributing to PC3. The horizontal lines represent the mean contribution (solid black line), the mean ± 1 standard deviation (dotted black line), and the mean ± 2 standard deviations (dotted green line). The 100 SNPs with the highest contribution are colored in red in the figure. Source: [32].

Species	Nubian ibex	Alpine ibex	Bezoar ibex	Taggar
Nubian ibex		0.41 | 0.85	0.33 | 0.90	0.28 | 0.46
Alpine ibex	0.41 | 0.85		0.47 | 0.91	0.36 | 0.46
Bezoar ibex	0.33 | 0.90	0.47 | 0.91		0.17 | 0.35
Taggar	0.28 | 0.46	0.36 | 0.46	0.17 | 0.35	
Combined [a]	0.14 | 0.28	0.20 | 0.23	0.10 | 0.32	0.03 | 0.06

[a]*Combined: the species mentioned in the column against the combined group of the other three species. Source: [32].*

Table 2.
Estimates of pairwise genetic differentiation (FST) between species (coded as A | B, where A is the underestimation of FST based on 33,698 SNPs and B is the overestimation based on 662 SNPs segregating in at least three out of four populations).

6.2 Goat breeds and wild goat species

The study included five domestic goat breeds from Sudan (Nubian, Desert, Nilotic, Taggar, and Saanen) and examined three wild goat species: Bezoar ibex (*Capra aegagrus aegagrus*), Nubian ibex (*Capra nubiana*), and Alpine ibex (*Capra ibex*).

6.3 High-density capture sequencing

Researchers employed a high-density capture sequencing approach to sequence the DNA of thirty-three goats. Through this method, they identified a total of twenty-two non-synonymous (amino acid-changing) and thirteen synonymous (non-amino acid-changing) single nucleotide polymorphisms (SNPs). Among these SNPs, nine non-synonymous and seven synonymous SNPs were previously unknown (**Figures 5–8**).

6.4 New casein protein variants

The identification of these DNA sequence variants led to the discovery of nine new casein protein variants. These variants were found in various goat breeds and wild goat species. For example, Alpha S1 casein variants were found in Saanen

Figure 5.
Capra nubiana-*male from Red Sea State, Sudan.*

Figure 6.
Capra ibex *from Leibniz Institute for Zoo and Wildlife Research Berlin, Germany.*

Figure 7.
*Male and female of Bezoar ibex (*Capra aegagrus aegagrus*). Source: http://www.shikarsafaris.com/en/anatolian-bezoar-ibex-turkey.*

Figure 8.
Taggar goats Capra hircus *from North Kordofan State, Sudan. Source: http://www.shikarsafaris.com/en/anatolian-bezoar-ibex-turkey.*

goats (CSN1S1* C1), Bezoar ibex (CSN1S1* J), and Alpine ibex (CSN1S1* K). Beta and kappa casein variants were identified in Alpine ibex (CSN2* F and CSN3* X). Alpha S2 casein variants were present in all domesticated and wild goats (CSN1S2* H), Nubian and Desert goats (CSN1S2* I), or Nubian ibex only (CSN1S2* J and CSN1S2* K).

6.5 Significance and conservation implications

The presence of most novel SNPs and protein variants was observed in the critically endangered Nubian ibex. This emphasizes the significance of conserving this endangered breed. Further validation and characterization of these new casein protein variants are necessary to enhance our understanding of their functional implications and potential [34].

7. The current conservation efforts for the Nubian ibex in Sudan

7.1 Status and importance

The Nubian ibex is currently classified as an endangered species on the IUCN Red List, indicating the urgent need for conservation efforts. The population of Nubian ibex has experienced a decline primarily due to various factors, including habitat loss, hunting, and the impacts of climate change.

7.2 Challenges and threats

Illegal hunting poses a significant threat to the survival of Nubian ibex populations. The indiscriminate killing of these animals for various purposes has contributed to their dwindling numbers. Additionally, human activities such as urbanization and infrastructure development have resulted in habitat degradation, further exacerbating the challenges faced by the Nubian ibex. Furthermore, the changing climate has disrupted the availability of suitable habitats and resources, posing additional threats to their survival.

7.3 Conservation measures

To address these challenges, several conservation measures have been implemented. Awareness campaigns have been conducted to educate local communities, tourists, and policymakers about the importance of protecting the Nubian ibex. Efforts have also been made to establish and manage protected areas where Nubian ibex populations occur, ensuring the preservation of their habitats. Strengthening law enforcement to prevent illegal hunting has been a crucial aspect of conservation efforts. Regular monitoring of population trends, health, and habitat conditions has been undertaken to gather valuable data for informed decision-making. Genetic management through breeding programs has been employed to maintain genetic diversity within captive populations. Additionally, translocations of individuals between populations have been considered to enhance genetic exchange. Education programs involving schools and universities have also been initiated to engage the younger generation in conservation efforts.

7.4 Collaboration and international cooperation

Recognizing the transboundary nature of the Nubian ibex populations, collaboration with neighboring countries such as Sudan and Egypt has been crucial in protecting these species. Working together with international organizations and wildlife institutes has also been instrumental in sharing knowledge, expertise, and resources to enhance conservation efforts.

7.5 Legal framework

To ensure the effective protection of the Nubian ibex, the implementation and enforcement of laws and regulations specifically aimed at safeguarding these species have been prioritized. Issuing licenses for conservation purposes and conducting annual surveys have been important tools in monitoring and managing the population.

7.6 Long-term vision

The long-term vision for Nubian ibex conservation is to develop comprehensive strategies that address all aspects of their protection. This includes not only addressing immediate threats but also considering long-term sustainability. Fostered cooperation among stakeholders is essential to ensure the effective implementation of the conservation plan [35].

7.7 Conservation implications

7.7.1 Relate genetic findings to Nubian ibex conservation efforts

Genetic research is essential for the conservation of the Nubian ibex. By assessing genetic diversity, researchers can identify unique traits and vulnerabilities within populations. Understanding population structure and relatedness helps prioritize conservation efforts, focusing on preserving advantageous genetic variations. Genetic information also guides captive breeding programs, preventing inbreeding and enhancing genetic diversity. Regular genetic monitoring informs management decisions, while collaboration with neighboring countries enhances overall resilience. Public awareness and education are crucial in highlighting the importance of genetic diversity for the long-term survival of the species.

7.7.2 Emphasize the need for collaboration and targeted strategies

7.7.2.1 Collaboration

Engaging diverse partners is crucial for effective conservation. This includes governmental bodies, local communities, NGOs, scientists, and international agencies. Working with neighboring countries is critical for protecting Nubian ibex populations that cross borders. Sharing expertise and resources strengthens conservation efforts.

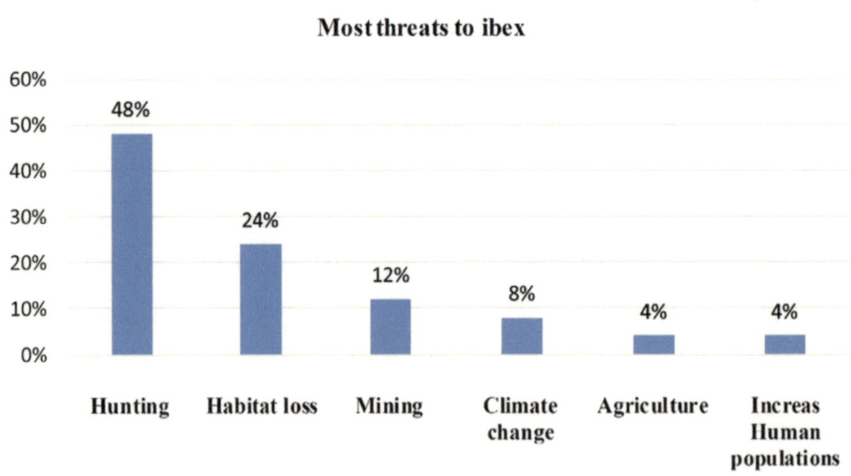

Figure 9.
Shows the main threats to Sudan's ibex population. Source: [35].

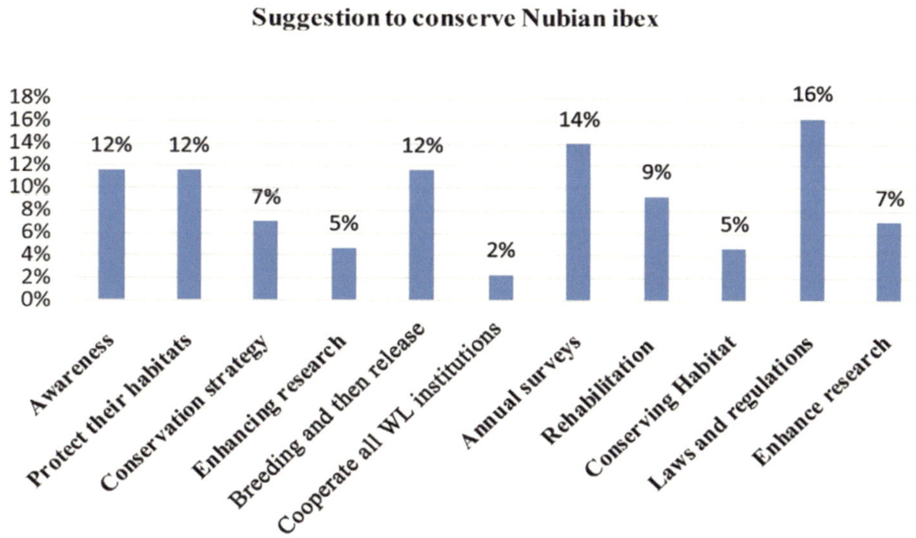

Figure 10.
Proposes a conservation strategy for Nubian ibex in Sudan. Source: [35].

7.7.2.2 Targeted strategies

Customizing approaches to address specific threats in each region is imperative. Prioritizing genetic preservation is crucial for maintaining diversity. Restoring critical habitats is essential for survival. Strategic anti-poaching measures based on population distribution and threat levels are key. Educational outreach to specific audiences raises awareness and support.

7.7.2.3 Comprehensive approach

Integrating expertise from genetics, ecology, and law enforcement is essential. Adapting strategies based on monitoring data and changing conditions is crucial for long-term success (**Figures 9** and **10**).

8. Conclusion

8.1 The key findings

The primary objective of the three studies was to examine the genetic diversity and population structure of the Nubian ibex (*Capra nubiana*) in comparison to other ibex species, such as the Alpine ibex, Bezoar ibex, and domesticated Taggar goats. The researchers utilized the 50 K goat SNP chip to genotype the samples.

In terms of observed heterozygosity, the Nubian ibex exhibited a range of 0.02–0.44, using either 33,698 SNPs or 662 SNPs. Similarly, the Alpine ibex showed a range of 0.01–0.38 and the Bezoar ibex ranged from 0.13 to 0.38. The domesticated Taggar goats displayed similar values of 0.40–0.41 using both SNP sets.

The pairwise F_{ST} values, which indicate differentiation between species, revealed that the Bezoar ibex had values ranging from 0.17 to 0.35 when compared to Taggar goats, 0.47 to 0.91 when compared to Alpine ibex, and 0.33 to 0.90 when compared to Nubian ibex.

Regarding molecular variance, approximately 74–78% of the variance was explained by differences between species, while the remaining variance was attributed to differences among individuals. The cluster analysis identified two distinct clusters: one consisting of Nubian and Alpine ibex, and the other comprising Taggar goats and Bezoar ibex. This analysis demonstrated a clear separation of all four breeds.

The researchers also identified specific genomic regions that contained known milk protein genes contributing to species differentiation. These regions were found through the analysis of capture sequencing for casein variants in the Nubian ibex. The objective of this study was to identify casein protein variants in the Nubian ibex and compare them with domesticated goats and other wild goat species. The findings revealed the detection of nine new casein protein variants, primarily in the alpha S1, beta, and kappa caseins. Interestingly, the critically endangered Nubian ibex exhibited the highest number of novel SNPs and variants.

In terms of the current conservation efforts for the Nubian ibex in Sudan, the species is listed as endangered on the IUCN Red List. The main challenges faced by the Nubian ibex include hunting, habitat loss, and insufficient population data. To address these challenges, proposed measures include.

8.2 Discuss future research directions

Future research directions for the Nubian ibex can be focused on genetic diversity, casein variants, and conservation efforts. By delving into fine-scale population genetics, researchers can uncover population structures, migration patterns, and gene flow within Nubian ibex populations. Functional genomics can shed light on how specific genetic variants influence adaptations such as thermoregulation and foraging behavior. Comparative genomics can provide insights by comparing Nubian ibex genomes with other wild goat species and domesticated goats to identify conserved regions and unique adaptations. Additionally, proteomic studies can investigate the functional implications of casein variants, while exploring their effects on milk composition, cheese properties, and nutritional value. Health implications of specific casein variants can also be explored to understand their correlation with disease resistance or susceptibility, as well as associations with overall health and fitness. Current conservation efforts for the Nubian ibex in Sudan can be enhanced through long-term monitoring of population trends, habitat conditions, and threats, as well as assessing the effectiveness of conservation measures. Research can also focus on investigating how climate change affects Nubian ibex habitats and developing adaptive strategies to mitigate climate-related risks. Community engagement with local communities can further support sustainable conservation efforts and assess the socio-economic impact of conservation initiatives. Ultimately, interdisciplinary research combining genetics, ecology, and socio-economic aspects will contribute to a deeper understanding of the Nubian ibex and inform targeted conservation strategies.

Acknowledgements

We extend our sincere gratitude to all the esteemed authors whose invaluable research findings have significantly contributed to the development of this chapter.

Additionally, we would like to express our heartfelt appreciation to my younger sister, Tarteel Hassan, for her unwavering support throughout this endeavor.

Conflict of interest

The authors affirm no conflict of interest that could influence the research findings.

Acronyms and abbreviations

C	Capra
IUCN	International Union for Conservation of Nature
WRC	Wildlife Research Center
WCGA	Wildlife Conservation General Administration
CITES	Convention on International Trade in Endangered Species of Wild Fauna and Flora
UNEP	United Nations Environment Programme
SNPs	single nucleotide polymorphisms
F_{ST}	Pairwise F_{ST} values for differentiation between species
PC	principal components
NGOs	non-governmental organizations

Author details

Lubna M.A. Hassan[1*] and Elhady A.M. Omer[2]

1 Wildlife Research Center, Animal Resource Research Corporation, Ministry of Animal Resources and Fisheries, Bahri, Sudan

2 Department of Animal breeding and Genetics, University of Khartoum, Khartoum, Sudan

*Address all correspondence to: lubnahamohamed@gmail.com
and lobnamoh2010@yahoo.com

References

[1] Ross S, Elalqamy H, Al Said T, Saltz D. *Capra nubiana*. The IUCN Red List of Threatened Species 2020: E.T3796A22143385. Gland, Switzerland: IUCN; 2020. DOI: 10.2305/IUCN.UK.2020-2.RLTS.T3796A22143385.en

[2] Habibi K. The Desert ibex: Life History, Ecology and Behaviour of the Nubian ibex in Saudi Arabia. London, United Kingdom: National Commission for Wildlife Conservation and Development (NCWCD); 1994

[3] Hakham E, Ritte U. Foraging pressure of the Nubian ibex Capra ibex nubiana and its effect on the indigenous vegetation of the En Gedi nature reserve, Israel. Biological Conservation. 1993;**63**(1):9-21. DOI: 10.1016/0006-3207(93)90068-C

[4] Tadesse SA, Kotler BP. Seasonal habitat use by Nubian ibex (*Capra nubiana*) evaluated with behavioral indicators. Israel Journal of Ecology and Evolution. 2011;**57**(3):223-246. DOI: 10.1560/IJEE.57.3.223

[5] Islam Z-U, Boug A, Judas J, As-Shehri A. Conservation challenges for the Arabian leopard (Panthera pardus nimr) in Western highlands of Arabia. Biodiversity. 2018;**19**(3):1-10. DOI: 10.1080/14888386.2018.1507008

[6] Alkon PU, Harding L, Jdeidi T, Masseti M, Nader I, de Smet K, et al. *Capra nubiana*. The IUCN Red List of Threatened Species 2008. Gland, Switzerland: IUCN; 2008. Available from: http://www.iucnredlist.org/details/3796/0

[7] Grubb P. Order Artiodactyla. In: Wilson DE, Reeder DM, editors. Mammal Species of the World: A Taxonomic and Geographic Reference. 3rd ed. Vol. 1/2. Baltimore, Maryland, United States; The Johns Hopkins University Press; 2005. pp. 637-722

[8] Stuart C, Stuart T. Mammals of North Africa and the Middle East. London, United Kingdom: Bloomsbury Publishing; 2016

[9] HCENR Higher Council for Environment and Natural Resources. Fifth National Report to the Convention on Biological Diversity (CBD). Sudan: Ministry of Environment, Forestry and Physical Development; 2014

[10] Damm GR, Franco N. CIC Caprinae Atlas of the World. CIC International Council for Game and Wildlife Conservation. Budakeszi, Hungary: Rowland Ward Publications; 2014

[11] Attum A, El Noby SK, Hassan I. The influence of landscape characteristics and anthropogenic factors on water hole use by vulnerable Nubian ibex *Capra nubiana*. Oryx. 2009;**43**(4):564-567

[12] El Alqamy HE, Wacher TJ, Hamada A, Rashad S. Camera traps: A non-invasive sampling technique to redefine the large mammal fauna of South Sinai. In: Full Book-2003. Muri, Switzerland: Cat Specialist Group-IUCN; 2003

[13] Rashad S. The Status of Nubian Ibex from 1999 to 2008 in Saint Katherine Protectorate. Saint Katherine Protectorate, Nature Conservation Sector, Egyptian Environmental Affairs Agency; 2008. Unpublished report, 28 p

[14] El Alqamy H, Ismael A, Abdelhameed A, Nagy A, Hamada A. Predicting the status and distribution of the Nubian ibex (*Capra nubiana*) in the high-altitude mountains of South Sinai (Egypt). Galemys. 2010;**22**:517-553

[15] Shkedy Y, Saltz D. Dispersal patterns and fragmentation of Nubian ibex and hyrax populations in Israel. Conservation Biology. 2000;**14**(1):200-206

[16] Royal Society for Conservation of Nature (RSCN). Annual Report 2015. Amman, Jordan: RSCN; 2015. 57 p

[17] Qarqaz M, Baker M. The Leopard in Jordan. Vol. 1. London, United Kingdom: Cat News; 2006. pp. 9-10

[18] Harrison DL. The Mammals of Arabia. London, UK: Ernest Benn Limited; 1968

[19] Eid E, Handal R. Illegal hunting in Jordan: Using social media to assess impacts on wildlife. Oryx. 2018;**52**(4):730-735

[20] Khorozyan I, Stanton D, Mohammed M, Al-Ra'il, W., & Pittet, M. Patterns of co-existence between humans and mammals in Yemen: Some species thrive while others are nearly extinct. Biodiversity and Conservation. 2014;**23**(8):1995-2013

[21] Barichievy C, Sheldon R, Wacher T, Llewellyn O, Al-Mutairy M, Alagaili A. Conservation in Saudi Arabia; moving from strategy to practice. Saudi Journal of Biological Sciences. 2018;**25**(2):290-292

[22] UNEP-WCMC. Report on Species/ Country Combinations Selected for Review by the Plants Committee Following CoP17. Cambridge: UNEP-WCMC; 2018

[23] Hillman JC, Yohannes H. Eritrea. In: Shackleton DM, editor. Wild Sheep and Goats and their Relatives. Status Survey and Conservation Action Plan for Caprinae. Gland, Switzerland: IUCN/ SSC Caprinae Specialist Group, IUCN; 1997. pp. 26-27

[24] Goodman I, Shachar N, Kahila G, Saltz D, Bar-David S. Evaluation of noninvasive genetic methods for Nubian ibex. Conservation Genetics Resources. 2017;**9**(2):181-183

[25] Saltz D, Berger-Tal O, Motro U, Shkedy Y, Raanan N. Conservation implications of habituation as determined by changes in the vigilance-group size effect of Nubian ibex in response to ecotourism. Animal Conservation. 2018;**22**(3):220-227

[26] Freiman E. Sex-specific population genetic structure: the Nubian ibex in the Judean desert [thesis]. Israel: Ben Gurion University; 2019

[27] Halevi T. Potential impact of human-dominated habitats on the population genetic structure of the Nubian ibex in Israel [thesis]. Israel: Ben-Gurion University; 2019

[28] Panthera. Arabian Leopard Initiatives: A Conservation Partnership. New York, United States: Panthera; Available from: https://www.panthera.org/new-life-leopards; 2019 [Accessed: February 23, 2020]

[29] Berger-Tal O, Saltz D. Invisible barriers: Anthropogenic impacts on inter- and intra-specific interactions as drivers of landscape-independent fragmentation. Philosophical Transactions of the Royal Society B: Biological Sciences. 2019;**374**(1781):20180049. DOI: 10.1098/ rstb.2018.0049

[30] Tosser-Klopp G, Bardou P, Bouchez O, Cabau C, Crooijmans R, et al. Correction: Design and characterization of a 52K SNP chip for goats. PLoS One. 2016;**11**(3):e0152632. DOI: 10.1371/ journal.pone.0152632

[31] Sayre B, Tosser-Klopp G. Design and characterization of a 52K SNP chip

for goats. PLoS One. 2014;**9**(1):e86227. DOI: 10.1371/journal.pone.0086227

[32] Hassan LMA, Arends D, Rahmatalla SA, Reissmann M, Reyer H, Wimmers K, et al. Genetic diversity of Nubian ibex in comparison to other ibex and domesticated goat species. European Journal of Wildlife Research. 2018;**64**:52. DOI: 10.1007/s10344-018-1212-z

[33] Rahmatalla SA, Arends D, Reissmann M, Ahmed AS, Wimmers K, Reyer H, et al. Whole genome population genetics analysis of Sudanese goats identifies regions harboring genes associated with major traits. BMC Genetics. 2017;**18**:92. DOI: 10.1186/s12863-017-0553-z

[34] Rahmatalla SA, Arends D, Said Ahmed A, Hassan LMA, Krebs S, Reissmann M, et al. Capture sequencing to explore and map rare casein variants in goats. Frontiers in Genetics. 2021;**12**:620253. DOI: 10.3389/fgene.2021.620253

[35] Hassan LMA, Adam MEA, Abakar LAA, Brema NYG. The current conservation efforts and future prospects for the endangered Nubian ibex (*Capra nubiana*) in Sudan. Journal of Environmental Protection. 2024;**15**(1):39-62. DOI: 10.4236/jep.2024.151004

Chapter 3

The Australian Endangered Pygmy Bluetongue Lizard and Livestock Grazing

Michael G. Gardner

Abstract

Endangered pygmy bluetongue lizards are now only found on livestock grazed properties in the Mid-North of South Australia, in a highly fragmented habitat of remaining native grasslands. This relationship has persisted since European colonisation of Australia. The lizards rely on abandoned trapdoor and wolf spider burrows for their homesites, and tilling of the soil is directly detrimental to both lizards and spiders. Overgrazing is also damaging to lizard survival and to burrow stability. Hence, pygmy bluetongue lizards are now reliant on farming practices that are consistent with the needs and survival of both the lizards and spiders. Fortunately, techniques such as rotational/planned grazing not only makes these lands more agriculturally productive they are also likely to be beneficial for this endangered species—although further research is required to ensure compatibility. Grazing management workshops have helped to spread the word about these new techniques and to the lizard's plight. Public support, especially those of graziers with this endangered species on their land, is vital for the continued survival of the species. Working in tandem, farm practices and endangered species management can be of benefit for property owners and endangered species alike.

Keywords: Egerniinae, Tiliquini, grassland, Squamata, Scincoidea, spider burrows, biodiversity

1. Introduction

Globally, grasslands are a large component of terrestrial environments, covering 25% of land surface [1]. The importance of grasslands for biodiversity is well documented [2, 3]. Despite their importance to food security and wildlife, grasslands are under threat with close to 50% already having been degraded worldwide [1, 3, 4]. The drivers of this degradation are varied and include processes such as invasive species and climate extremes confounding altered fire regimes as well as use changes such as land conversion to forestry, crops, or housing [2, 4]. Other damaging factors fall into agricultural mismanagement such as over grazing, eutrophication, and land abandonment [2, 4]. These issues are compounding the threat, especially given most of the earth's natural grasslands now lie within agriculturally managed areas [3].

Intensively managed grasslands are often thought to have less conservation importance than grasslands in their native context [5]. Whilst floristically these areas may well be less intrinsically valuable, agriculturally managed grasslands can be vital for supporting fauna when the presence of intact high quality natural grasslands is limited or non-existent [3, 5]. Mostly, large herbivores such as bison, antelopes etc. and their predators, wolves and lions for example, come to mind when one considers the endangered species supported by agricultural grasslands. However, agricultural grasslands also are home to many smaller and less known creatures, with some species, such as the black footed ferret (*Mustela nigripes*), having been brought back from near extinction. Some endangered species, for example the plains-wanderer (*Pedionomus torquatus*), live with some level of reliance on grazing of agricultural land as an important component of their survival [6, 7]. A similar reliance on livestock grazing management occurs with the endangered South Australian endemic pygmy bluetongue lizard (*Tiliqua adelaidensis*) [8, 9]. This species exists in remnants of untilled land only in the Mid-North region of South Australia (**Figure 1**). All known natural populations of this lizard are found on livestock grazed properties (**Figures 1** and **2**) on the ancestral lands of the Kaurna, Nukunu, Narungga and Ngadjuri people [10].

Despite agriculture and biodiversity seemingly having end goal differences, livestock managed grasslands can provide a valuable role in the maintenance of biodiversity [2, 5–7, 11]. Here I discuss the intimate relationship that the endangered pygmy

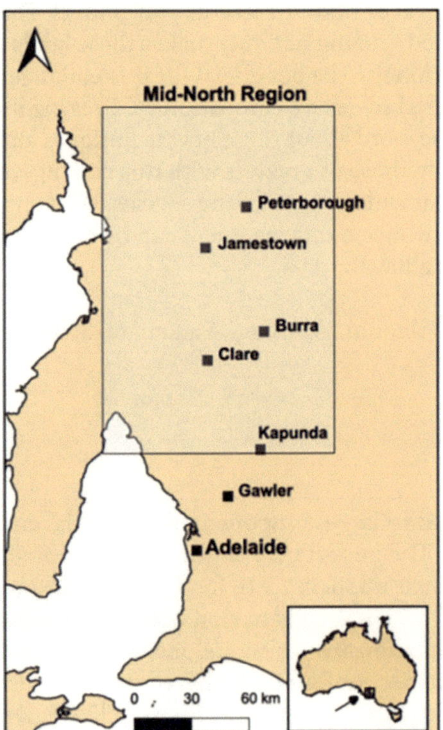

Figure 1.
A map of the area known as the Mid-North of South Australia with towns/cities relevant to the pygmy bluetongue lizard's distribution marked.

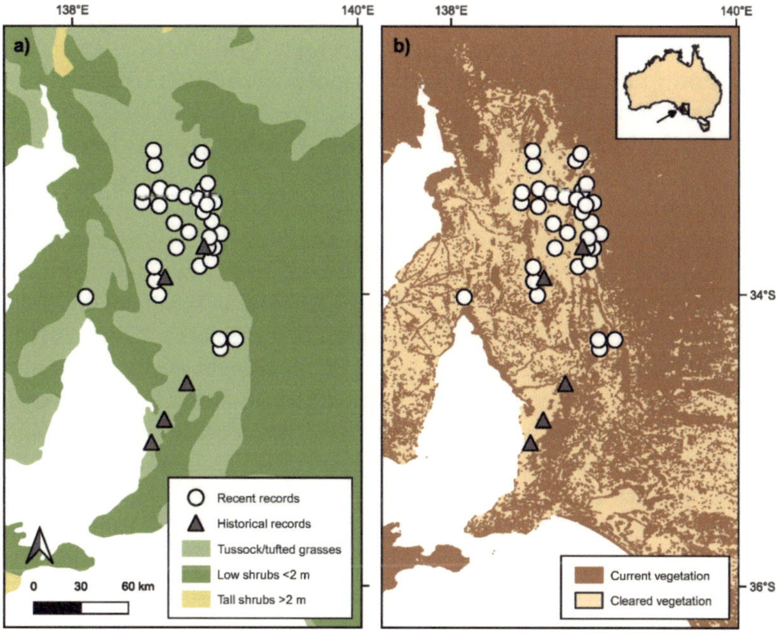

Figure 2.
(a) Map overview of recent and historical records of pygmy bluetongue lizards and the pre-European dominant vegetation type of the lower stratum, and (b) Map of the extent of cleared vegetation in the Mid-North region of South Australia and the relationship to pygmy bluetongue distribution. Mapping layers from Nature Maps.

bluetongue has with agriculturally managed grasslands. Within this context, the fate of the pygmy bluetongue is inextricably linked to the current and future maintenance of these grasslands—and the species would likely go extinct if livestock was removed without other compatible management systems in their place [9].

2. Grasslands in the mid-north of South Australia

South Australia's Mid-North is north of the Adelaide Plains, extending as far south as Kapunda at the northern end of the Mount Lofty Ranges, west to the Spencer Gulf, north to the Southern Flinders Ranges, and east to Robertstown (**Figure 1**). This region has a Mediterranean climate with plant growth limited by temperature in winter and dryness in summer, and is in the Mount Lofty Block bioregion [12]. The area's current vegetation is normally classified as 'temperate grasslands' under the Köppen classification scheme [5].

South Australia has undergone substantial modification since European settlement (**Figure 2b**). Settlers likely occupied much of the temperate grasslands within the Mid-North by the 1860's and 1870's [13]. For example, Kapunda (**Figure 1**), a town at the most southward extent of pygmy bluetongue current occupation, was settled by pastoralists in 1839—only 3 years after the European colonisation of South Australia [13].

2.1 Pre-European vegetation in the mid-north of South Australia

Early after European colonisation, there are accounts of large 'swaths' of Kangaroo Grass (*Themeda australis*) in the Adelaide plains [14]. Moore [14] suggested that after substantial—primarily sheep—grazing, the kangaroo grass plains in the Adelaide plain gave way to tussock grasslands dominated by scented "Iron-grass" *Lomandra effusa* and/or Stiff "Iron-grass" *Lomandra multiflora* subsp. *dura* (both from the Family Asparagaceae) with a wide variety of introduced species such as wild oats (*Avena barbota*) [14]. Early European reports for the Mid-North suggest at least some areas, such as around Gawler in the south of the Mid-North region, also consisted of diverse grasslands [15] with early European paintings—particularly those by George French Angas—showing very open park-like landscapes in the Mid-North [16]. Reports by colonists indicate that the areas around Burra were 'largely devoid of vegetation', interpreted as being grasslands, and that the areas south of Templers to Gawler were a 'cheerless' plain, possibly indicating the occurrence of tussock grasslands (**Figure 2a**) [17]. Jessup 1948 contends that prior to European settlement and subsequent cropping and general habitat changes, at least parts of the Mid-North were essentially treeless and likely were grasslands [17]. Therefore, most consider that the Mid-North at the time of European colonisation was largely a native grassland [18].

The current grassland type, iron-grass natural temperate grassland, is thought to be unique and endemic to South Australia [19] and it has a Recovery Plan to support the recovery and stop the decline of plants within this unique community [19]. The existence of mature tussocks is an indication that the land has not been tilled and supports species sensitive to cultivation and fertilisers [19]. As such the natural grasslands are an important refuge for such species such as spiders, soil-dwelling insects and reptiles such as the pygmy bluetongue lizard [19]. The extent of the grassland, once thought to be between 750,000 to 1,000,000 hectares [20], is now much reduced and in 1995 was estimated to be less than 1% of its original distribution (**Figure 2b**) [21].

2.2 The pre-European maintenance of grasslands

Grasslands worldwide have been maintained globally by a combination of fire and grazing and had their maximum extent in the Holocene (~8000–4000 years BP) [22]. The origin and maintenance specifically of Australian grasslands has been controversial, with limited ecological research to understand the role of Aboriginal people in the maintenance of grasslands via fire [23–25]. The evidence from palaeoecology and palynology is less clear than that of historical accounts, with most scientific accounts suggesting there is only limited ecological evidence for firestick farming to be responsible for creating and maintaining grasslands [23]. Actual evidence for fire-stick farming within the Mid-North is lacking.

The grasslands of the Mid-North, regardless of their previous maintenance, are now primarily managed as livestock feeding areas for sheep [26] although some cattle grazed properties exist in the area. Before the early 1800's, the grasslands supported mobs of kangaroos [18], which are still found within the area although likely in lower numbers due the sustainable management of commercially harvested kangaroos [27].

3. The endangered pygmy bluetongue

The pygmy bluetongue lizard (*Tiliqua adelaidensis*; **Figure 3a**), was rediscovered in 1992 after being thought extinct for a little over 30 years [28]. A detailed account of the rediscovery and biology of the species has been provided previously [9] however some background is provided here for context. The species is endangered (IUCN

Figure 3.
(a) An adult pygmy bluetongue lizard, and (b) Typical pygmy bluetongue habitat in Burra during the austral spring. Photos by author.

red list) [29] and was listed in Appendix 1 of Convention on International Trade in Endangered species (CITES) at the Conference of the Parties in 2022 (CoP19). The species only occurs naturally in highly fragmented grasslands of the Mid-North of South Australia (**Figure 2b**) [8, 9]. Currently there are ~38 sites where these lizards have been found that are at least 2 km from other occurrences. Most are found on sheep grazed grasslands with two populations found on cattle farms. No natural populations exist outside areas managed for livestock although one site is under the care of the Nature Foundation, a not-for-profit organisation which is South Australia's largest and oldest private land conservation foundation. This property is managed as a sheep grazing area.

3.1 An unusual home site explains limited early records

An extraordinary lifestyle provides some explanation for the limited early observations—only 20 up until 1959 [9]—and listing as possibly extinct in a popular herpetofauna guide published in 1992 [30]. This medium sized skink spends up to 98% of its time within spider burrows, specifically those of lycosid (wolf spiders) and mygalomorph (trapdoor) species [8, 31]. These spiders dig holes that, when abandoned, are used by the lizards [31–33] creating an interdependency between elements of grasslands as found in other systems such as the one existing between black footed ferrets and their prairie dog (*Cynomys* spp.) prey [34]. Even more similar is the relationship of earless dragons (*Tympanocryptis* spp.) with arthropod burrows [35].

Pygmy bluetongue lizards do not dig their own burrows [31] and much less research has been focused on the spiders despite their importance to the survival of the pygmy bluetongue (but see [32]). The fate of the spiders, whose burrows the lizards use, is also intrinsically linked to the widespread agricultural practices in the last nearly 250 years [36]. The tilling of soil and associated compaction mechanically disrupts ground digging spiders and their burrows and also disrupts soil structure such that the ground is not suitable for an extended period — perhaps decadal disruption occurs [9, 37].

3.2 The relationship between grasslands and the pygmy bluetongue

Much of the Mid-North was tilled for cropping post European settlement, with parts of South Australia some of the most cleared landscapes in Australia [38]. The result of the conversion of the grasslands to cropping is the fragmentation of potentially suitable spider and pygmy bluetongue habitat. Remaining pygmy bluetongue habitat occurs only in areas that were too steep to warrant tilling, such as those surrounding hills, or in areas too dry to be considered viable for crops (**Figure 2b** and **3b**). Likely the areas where the lizards survive represent only marginal habitat with all prime habitat areas made unsuitable by either infrastructure or tilling [9]. As stated previously, the remaining suitable semi-natural grassland habitats are all grazed, primarily by sheep. The grazing on these properties is likely to have occurred for over a century based on historical reports [14], so clearly the lizards are able to tolerate some level of grazing [9].

Vital to the pygmy bluetongue's survival then, are farming practices that are not detrimental or at least only minimally detrimental or neutral. Overtime, practices should become sensitive to the plight of the lizards as appreciation of the lizards and their habitat add additional motivation. As documented by [9] the species has been under intensive research since its rediscovery. Early experimental research, to understand the relationship of the lizards to grazing levels around their burrows, simulated heavy grazing—defined as clipping grass to the ground level without any associated heavy trampling—in both field and laboratory situations [39]. Although the lab experiments were equivocal, field experiments established a clear preference for burrows with vegetation, with no lizards settling into burrows in the simulated grazing areas [39]. These results complimented research showing that lizards would not colonise tilled fields [40] even with the addition of artificial burrows, which they will readily use [41]. Other grazing experiments also supported the negative aspects of grazing, with lizards having lowered body mass increases with simulated grazing compared to controls of no grazing [42]. The treatments also increased the tendency of lizards to move away from grazed patches [43]. Taken together, these studies show a need for some level of cover provided by the presence of grass around burrows. Other recent research suggests that tussock grass is an important resource for temporary basking refuges and for foraging opportunities [44]. Grass cover would also be important during spring when males move to find mating opportunities [45, 46]. There may also be positive thermal aspects of the presence of vegetation. It seems clear that some level of vegetation is important for the lizards.

Spider burrow persistence and construction of new spider burrows is also linked to the presence of a native grassland and an undisturbed soil layer. Within one austral spring/summer period, these aspects of spider activity were compared in ungrazed and grazed plots by Clayton et al. [47]. Both spider recruitment and persistence of wolf spiders were negatively affected by grazing, although the abundance of trapdoor spiders was not affected. The burrows of trapdoors spiders are favoured by lizards, above wolf spiders burrows, due to their greater depth [31] and stability [47] with most larger trapdoor burrows occupied by either spiders or lizards. Lizards do not remove spiders from their burrows—even though lizards may be able to kill the spiders, they are also able to be killed by them [48]—and instead, burrows occupied by lizards have been found to be unoccupied in the month before the lizard moved in [32]. There may be some as yet unexplored positive stability aspect of burrow occupancy as lizard occupied burrows have been found to preferentially persist longer than unoccupied burrows [49] even in ungrazed plots. The greater persistence of burrows in ungrazed vs. grazed plots was also confirmed [49] similarly to [47] where the number of burrows were shown to have increased in ungrazed plots. The later study also reported, in grazed plots, a loss of shallower burrows and those with a smaller diameter than usable for lizards [47].

It is likely therefore that grasses provide stability for burrows, and this aspect was explored opportunistically after a rain event flooded burrows. A much larger proportion of burrows not surrounded by vegetation were destroyed [50] indicating a longer term positive aspect of grass around burrows.

In contrast to studies showing the detriments of grazing, prey capture ability by lizards may be enhanced by grazing. Pettigrew et al. [51, 52] showed that lizards made more successful prey attempts within experimental grazing treatments. Perhaps this is due to grazed areas providing greater visibility for the sit-and-wait predation method used by these lizards [9]. A caveat is that the research was done during drought years which may not be reflective of the situation in higher rainfall seasons [51]. Additionally, the basking opportunities of lizards were enhanced in these experimentally grazed plots as lizards basked fully emerged, with proportionally more of their bodies exposed on the surface [51]. Nielson et al. [53] showed, with data from three sheep grazing treatments (high, medium and low), that irrespective of treatment, lizards were found in locations with less vegetation cover than random, confirming this benefit of increased basking. In another aspect of this same study, grasshoppers, a common food item of the lizard, increased with decreasing grazing intensity [42]. A complex association of lizard body condition with grazing was found, and this differed between the two years of the study. Within the first year, lizards in ungrazed paddocks had greater body condition and yet, in the second year, those in the intermediately grazed treatment were higher [42]. Clearly there are distinctions that short time frame studies are unable to elucidate, and more research is needed over longer time periods.

There does, therefore, appear to be a nuanced relationship to the amount of grass cover that is beneficial to lizards. Recent research [54] shows that lizards exhibit microhabitat preferences within the grasslands. There was a tendency for lizards to occupy burrows that are closer to grass tussocks—supporting previous work indicating that some level of cover may provide foraging and basking benefits to lizards [43–46]. This recent study [54] was conducted in high rainfall years, and further research is required to understand if there is a switching of preferences during different climate systems with lizards potentially preferring open burrows during the wetter La Nina cycles and then burrows surrounded by greater vegetation under El Nino years when the predominate climate in the Mid-North is drier.

Hence, understanding the level of grazing that is of most benefit to the lizards longer term is an important research focus area. Given the previous findings of both positive and negatives aspects of grazing, on balance, one could predict that an intermediate level of grazing might be most beneficial to lizards [9]. Further research should address the levels of grazing that promote long term persistence of pygmy bluetongue populations, comparing levels of grazing that are common in the Mid-North of SA with alternatives that are designed to provide greater livestock yields to landowners.

4. Grazing management for the survival of the pygmy bluetongue

4.1 Various grazing management techniques

4.1.1 Traditional grazing

In set stocking, also called continuous grazing [26, 55], stock are kept within the same paddocks for extend periods such as between 3 and 12 months (**Figure 4**).

Figure 4.
Pictorial of set stocking where livestock are kept within the same paddock for long periods of time. Adapted from [56]. Drawn using Biorender.com.

Numbers of animals may be managed to suit particular amounts of feed, but grazing is continuous and the plants do not have a recovery period during crucial stages of growth so some native pastures will not tolerate this grazing method [56, 57]. Set stocking has been shown to be detrimental to biodiversity [58].

4.1.2 Rotational grazing

One way to provide intermediate levels of grazing that is also beneficial to sheep productivity [18, 26] is that of 'rotational grazing'—although some contest the financial benefit [55]. In contrast to set stocking, rotational grazing (**Figure 5a**) involves having between five to seven paddocks, according to [53], or between as few as two to four paddocks [26] available to a mob of sheep. The mob are then moved through the paddocks after a certain number of days regardless of the pasture condition (e.g. every 10 days) or based on stock condition—if it is noticed that the stock are losing condition for example [53].

A variation on rotational grazing is 'planned grazing' (**Figure 5b**)—also known as cell grazing or high density, short duration grazing—where the stock are moved between paddocks according to plant growth [56]. In this case, the paddock is grazed with as many stock as possible for a short period of time (e.g. 1–3 days) before stock are moved on to the next paddock, spending different numbers of days within each paddock according to the paddock size, plant growth and biomass. Planned grazing trails have shown the greatest pasture productivity, highest stocking rates, as well as the least bare ground [56]. Native Australian plants do well under this management practice with perennial grasses increasing their size and health [56] and there is evidence that planned grazing does promote plant diversity [59]. In North America, research shows a reduced stock performance benefit with heavy stock densities and promotes a moderate density as being most beneficial [55]. Workshops are held within the Mid-North Region by members of the pygmy bluetongue recovery team to educate graziers of the benefits of these farming techniques for the united goals of profitability and lizard conservation. Positive public perception is key to the continued survival of the species. People need to care enough to make changes to their practice, although some may be convinced on the potential for profitability increases alone [18]. Other landowners may be more sympathetic to conservation and will be intrinsically motivated [18].

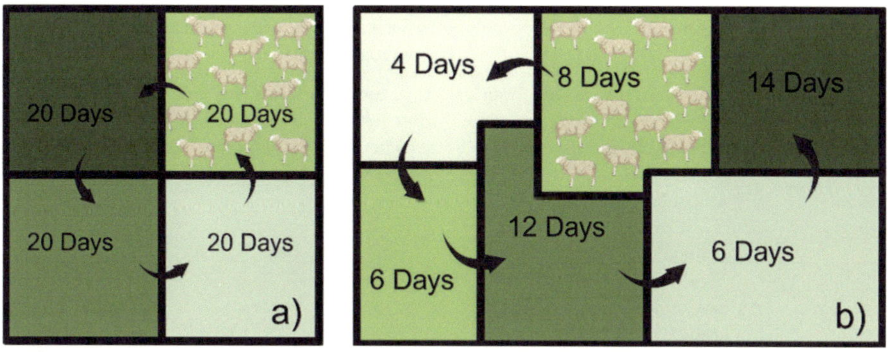

Figure 5.
(a) Pictorial of rotational grazing where livestock is moved between paddocks following a set time period, here 20 days, or after a loss of body condition is noticed. Each paddock is given a similar time frame to recover with recovery represented by cell shading, and (b) A pictorial of planned grazing, a variant of rotational grazing where stock are moved according to plant growth (represented by shading) and paddock size, spending longer in paddocks where plant growth is faster. Paddocks have time to recover under both scenarios. Adapted from [56]. Drawn using Biorender.com.

4.1.3 Principals of planned grazing and potential impacts on lizards and spiders

The 'How to make money out of grasslands' document [56], provided to landowners as part of the workshops, indicates there are six principles of planned grazing management (**Table 1**). These principals have potential consequences to pygmy bluetongue and spider conservation, as indicated in **Table 1**. Generally, planned grazing could have beneficial impacts to both lizards and spiders but requires testing. These benefits include greater burrow stability due to less persistent trampling by livestock, more consistent prey availability with more consistent vegetation presence, limited periods of exposure to predators such as snakes and particularly birds, [60] and greater diversity of vegetation potentially increasing prey diversity (**Table 1**).

Regeneration of pastures relies on allowing plants to set seed every couple of years to provide growth of new plants [5]. Grazing needs to occur to keep the pasture healthy [11]. However, it is important to keep the roots of grazed plants strong by allowing time between grazing for root recovery [59]. There is not enough root material to support new leaf growth if the pasture is re-grazed during the recovery phase when the plants need to draw from the nutrients of roots to re-grow leaves. Therefore, recovery periods are the most important part of successful grazing management. The document recommends not taking more than half the dry matter out of pastures during each grazing event [56].

Aside from potential benefits to the lizards and spiders, rotational grazing should promote a biodiverse grassland [5, 11, 59] although landscape heterogeneity should be maximised for this affect to be realised [57]. Restricting the stock's time in a field is likely to reduce the mechanical degradation of vegetation and promote perennials [57]. The ideal diversity of perennial grass for a healthy grassland in the Mid-North is suggested to be greater than seven species, maximising species with different growth patterns [56]. Increasing the diversity to cover both annuals—that grow rapidly and provide a feed for stock in good times, but also perennials to get stock through lean times—would potentially also benefit lizards by providing constant cover (**Table 1**).

Pygmy bluetongues do not appear to have a preference for a particular vegetation community [61] and likely structure is more important so maximising species

Principle	Key points	Potential impact on pygmy bluetongues and ground dwelling spiders in the system
1. Ensure that plants get an adequate recovery period	Plants need time to recuperate and sufficiently re-grow to allow full root growth before being grazed again. An assessment of the amount of leaf material removed and an understanding of the growth rates of plants is required to determine the rest period needed.	Time between grazing may promote the accumulation of prey items in the pasture and may keep prey numbers at higher levels for longer periods, benefiting lizards and spiders.
2. Match stocking rates to carrying capacity	Carrying capacity of the land needs to be worked out to provide stocking rates. Utilisation of only 50% of standing feed is recommended.	This principal may provide consistency across different years and seasons in the amount of cover provided by grasses, with short time periods when cover will be less. Further research is required to understand if 50% utilisation of feed provides lizards with enough basking opportunities. Differences may occur in La Nina verses El Nino years which needs further research.
3. Utilise a short graze period	A short graze period of 3–4 days is preferable, with the number of paddocks being a limiting feature here.	A short but intense grazing may allow burrows to persist as stock trampling would be at times when there is good level of grass with strong roots, which would provide greater stability of burrows.
4. Ensure maximum stock density for the minimum time	A high stock density reduces selective grazing and encourages more even consumption of pasture plants, reducing the growth of less palatable plants.	Greater structure in the pasture would favour a larger number of tussock grasses that are important for the lizards. Perennials are likely to be less impacted providing continued cover and increased soil, and hence spider burrow stability.
5. Ensure livestock has an adequate water supply	Clean troughs and tanks are preferred rather than a dam. Water needs to be always available.	The use of troughs rather than dams would lead to a larger land footprint being available for spiders and lizards alike.
6. Plan, monitor and manage the whole grazing system	Use of the condition of the pasture rather than the condition of the stock for managing.	A well-managed system will lead to a more consistent environment. Research needs to be undertaken to ensure there are benefits for the lizards and spiders.

Table 1.
Principals of planned grazing management with their key points for management [56] and the potential impact on pygmy bluetongue lizards and spiders. Stocking rate is the number of animals on an area (usually one hectare) for a period of time (usually one year) and is measured in Dry Sheep Equivalents (DSE). Carrying capacity is a measure of how much vegetation a property can dependably produce for an infinite time.

diversity could well be important to provide greater structure. Broad-leaved plants are a good source of minerals, although they are less tolerant of grazing, but could be an important part of a diverse pasture [56]. Having more broad-leaved plants would provide structural cover for the lizards when they are out of their burrows, as well as likely promoting greater prey numbers and keeping soils together, promoting spider colonisation and burrow building (**Table 1**) [47]. A more diverse grassland is likely to be beneficial to stock and the pygmy bluetongue.

The type of management that is good for stock productivity is therefore likely to be good for lizards and the reverse is also likely to be true [9]. Stock animals will obtain most nutrition feeding when the plants are mature but still actively growing [62] as although young plants have the best levels of proteins and water, they have less energy with the energy levels in plants increasing as they mature. Livestock feeding when plants are mature and then moved on would promote burrow stability [47, 63]. In contrast, plants that have dried off have low levels of protein and do not provide enough to meet stock's nutrient needs [18] as dry pasture breaks down slower than high quality forage due to its woody nature and deficiency of nutrients [56]. If grass is eaten during times when it has dried off, not only are the plants less nutritious, but plant root growth will also be negatively impacted [39, 47, 63] with longer term consequences of poor plant recovery. The stability of burrows would be reduced if feeding happens at rates or times that have this negative impact on root growth (**Table 1**). Another negative consequence for the lizards would be the reduction of vegetation cover which likely could reduce lizard movement due to the perceived threat of predation when vegetation cover is sparse [51] and, if lizards do move, they are less likely to survive those moves. The result would be further reduction in gene flow across populations in a species which has limited dispersal [45, 64].

Further research should target understanding the longer-term benefits of rotational/planned grazing on the survival of the pygmy bluetongue. Additionally, little is currently known of the effects of seasonal differences in grazing regimes on pygmy bluetongues.

4.1.4 The unknowns of existing grazing management variations

The Mid-North graziers use either primarily summer or winter grazing in paddocks, with the regime even variable among paddocks within the same property. These regimes would promote variation in the plant communities found at seasonally differently grazed sites. The broad grass types, C3 and C4, would be encouraged and respond at different times of year. C3 plants have their highest nutritional value in winter and suit winter grazing regimes whereas C4 plants have their highest nutritional value in summer [65]. No studies have determined the effect of summer vs. winter grazing on the pygmy bluetongues either in the short or long term. Longer grass in spring may permit greater survival of lizards when they move, particularly relevant for males as they are known to leave burrows to seek mating opportunities in the austral spring [45]. Increased cover therefore potentially facilitates lizard movement and survival [43, 51, 66] and would likely have a positive consequence on the maintenance of genetic diversity in this promiscuously mating species [46]. However, it is unclear if the laying of scent trails [67] may be more difficult for females which could reduce mating opportunities. Although no research has established that females need bare ground to lay their trails, the presence of copious amounts of vegetation might hinder the ability to do so. Further research is required to understand the pros and cons of grazing in different seasons.

As stated, varied grazing regimes are found throughout the Mid-North among, and even within, properties with pygmy bluetongues. A salient point then, that the pygmy bluetongue recovery team makes to graziers with these lizards on their properties, is to keep with the grassland management that they have been doing [9]. Clearly what these property owners are doing has been compatible with the species survival. Of course, it might be these populations could be doing even better and with a further understanding of the main driving factors for species survival, the recovery team could advise for a better conservation outcome [18]. One potential barrier to conservation of the species is the attitude of some property owners. Graziers' memories are long—even generational—and interactions between with graziers and the recovery team needs to be positive and respectful with a view to the long term [9].

5. The future for the pygmy bluetongue under climate change

Finding sympathetic grazing regimes and graziers who are willing to work with the recovery team to promote the species survival is only one thrust of the species' future [9]. Areas in the north of the lizard's range are also under threat from climate change [68–70]. Due to the lizard's limited dispersal capacity [45, 64], areas of grassland are too far apart for any natural dispersal to connect them, and therefore any remaining areas further south cannot be occupied by relying on natural dispersal alone [64]. Modelling of various management options has suggested that regular and systematic relocation of lizards from declining populations to sites with high current habitat quality is the best option to maintain the species [69]. The models suggest assisted colonisation will be a vital component of any future management to prevent significant range contraction, or even extinction. Trial translocations are underway at an area outside the current range of the species, but at the same latitude as the southernmost population (Kapunda). Early results suggest that limited behavioural plasticity is a potentially limiting factor for moving lizards [71]. Lizards moved from a mid-northern population (Burra) showed greater daily activity and were active at higher temperatures and lower humidity than those moved from the southern edge of the current range (Kapunda). These results indicate that local adaptation within different populations may have occurred, and further research is required to understand the extent of behavioural plasticity and if the lizards from different populations will be able to survive at more southerly locations if further translocations are undertaken [71]. Understanding any genetic components of behaviour will also be important.

Understanding population level behaviours and the responses to different microclimatic conditions at a translocation site have implications for hydro-regulation, and when selecting individuals for assisted colonisation [71]. Further research will determine how long it might take for these animals to adjust their behaviours to new sites and if any differences are generational. For most species there is limited background information and time to test variables associated with translocation success. However, for the pygmy bluetongue, we have substantial information generated from previous research to underpin the strategies for undertaking assisted colonisation. We also have the luxury of time, as the species is under medium/long term threat from climate change rather than immediate threat [9].

That lizards may take time to adapt to new conditions is one important aspect to consider, but even finding land appropriate for translocations will be problematic. The large scale clearing in South Australia [38] means that most arable land has been used for either cropping or, increasingly, for dwellings or infrastructure. Only small areas are potentially available [70] and this is possibly a limiting factor for the future survival of the species. Regenerating arable soils into grasslands may be an option [37] however it would take time (decades?) for the soil structure to be suitable for spider reestablishment [9]. Although some properties that support populations of lizards are on previously tilled land, these are thought to have had decades of no tilling. Further research is required to understand if regenerating areas for future translocations is viable. Considerations also need to be made to understand if translocating spiders is part of the solution—clearly their conservation is also important [9].

In the more immediate future, agricultural or construction developments such as wind farms are being proposed in pygmy bluetongue habitat [9]. Ironically an activity to help gaining energy from sustainable sources, often touted as part of a solution [72], may be in direct collision with pygmy bluetongue conservation—the areas suitable for windfarms in the Mid-North are often co-located with pygmy bluetongue populations. Understanding how to balance the needs to make human habitation more climate change resistant with the possible impact on this endangered species requires further research.

6. Conclusion

Grasslands are a valuable resource worldwide for securing food for a growing world [26] yet are dwindling [2, 3]. These same grasslands can often be rich in biodiversity and are invaluable in providing habitat for grassland dependant species [3]. Species interdependencies occur within these systems such as with the pygmy bluetongue and the spiders it depends on, and other examples—such as exist between black-footed ferrets and their prey the prairie dog [34]—highlight that a whole ecosystem approach is needed [2]. Sometimes all that is left is a highly fragmented system, and, in the Mid-North of South Australia, these remaining grasslands are the only available critical habitat for the pygmy bluetongue lizard [9]. The long-term survival of this IUCN red listed endangered species is intrinsically linked to the sympathetic management of livestock within these grasslands [18].

Moves towards management compatible with lizard survival are being introduced through initiatives from state and federally funded projects, and these are also able to support a profitable livestock industry. Biodiversity is now being promoted by organisations such as the Australian Meat and Livestock Association [73]. With the pygmy bluetongue, the marriage of livestock production and endangered species management is a rare example of a relationship where seemingly opposed stakeholders can work in unison towards an ideal end result for both parties—rather than both having to compromise. More research, however, is needed to understand the longer-term implications of grazing regimes. On the surface, rotational/planned grazing appears to satisfy what the research thus far indicates are the lizards' grazing requirements—that intermediate levels of grazing are most compatible with lizard survival [9]. However, most research to date has taken place on the three-year cycles of grant funding. Longer term trends need to be elucidated to provide certainty going forward.

Nuances exist in grazing regimes at different properties where this species occurs. No research has examined the potential differences between summer and winter grazing and how these regimes impact on the species and only anecdotal observations around lizard movement exist. We can derive a hypothesis that summer grazing might result in a more connected genetic population compared to winter grazing, due to greater vegetation cover during the spring mating season, but this requires testing. Additionally, two properties with pygmy bluetongues are cattle farms and it is an open question if sheep or cattle are more compatible with lizard survival.

The cooperation of landowners is paramount to ensure a positive future for this endangered species. Bull and Hutchinson [9] make a salient point that community involvement, such as via The Pygmy Bluetongue Conservation Association, are key to the future of the species and associated grasslands. Worldwide there are other examples where grasslands are valuable habitat for biodiversity and are increasingly being seen as important conservation areas [2, 3, 5, 15, 57, 58, 74] and attention should be paid to maintaining and developing positive relationships with graziers to be able to satisfy conservation and economic outcomes.

Acknowledgements

I would like to thank the Australian Research Council (LP190100071) and a Saving Native Species (Priority Species SNSPA000008) grant from the Australian Federal Government for funding that allowed this chapter to be written. Emily Hoffman is thanked for preparing the maps using NatureMap layers https://data.environment.sa.gov.au/NatureMaps/Pages/default.aspx. I greatly appreciated critical comments on a draft of this chapter by Mark Hutchinson which helped to tighten the content. This work would not have been possible without the dedication of the late C. Michael Bull and his past students as well as my current students. C. Michael Bull had a systematic approach to science which laid the foundation for much of our knowledge on this endangered species. Previous funding has come from Australian Research Council grants to Bull and Hutchinson.

Conflict of interest

The author declares no conflict of interest.

Notes/thanks/other declarations

Some figures were drawn using Biorender.com.

Author details

Michael G. Gardner
College of Science and Engineering, Flinders University,
Bedford Park, South Australia, Australia

*Address all correspondence to: michael.gardner@flinders.edu.au

References

[1] Gang C, Zhou W, Chen Y, Wang Z, Sun Z, Li J, et al. Quantitative assessment of the contributions of climate change and human activities on global grassland degradation. Environmental Earth Sciences. 2014;**72**(11):4273-4282

[2] Bardgett RD, Bullock JM, Lavorel S, Manning P, Schaffner U, Ostle N, et al. Combatting global grassland degradation. Nature Reviews Earth and Environment. 2021;**2**(10):720-735

[3] Petermann JS, Buzhdygan OY. Grassland biodiversity. Current Biology. 2021;**31**(19):R1195-RR201

[4] Gibbs HK, Salmon JM. Mapping the world's degraded lands. Applied Geography. 2015;**57**:12-21

[5] Dorrough J, Yen A, Turner V, Clark S, Crosthwaite J, Hirth J. Livestock grazing management and biodiversity conservation in Australian temperate grassy landscapes. Australian Journal of Agricultural Research. 2004;**55**(3):279-295

[6] Nugent DT, Baker-Gabb DJ, Antos M, Collins L, Green P, Morgan JW. Livestock production land and conservation areas play a complementary role in the conservation of a critically endangered grassland bird. Ecological Solutions and Evidence. 2022;**3**(4):e12182

[7] Eriksson O. The importance of traditional agricultural landscapes for preventing species extinctions. Biodiversity and Conservation. 2021;**30**(5):1341-1357

[8] Hutchinson MN, Milne T, Croft T. Redescription and ecological notes on the pygmy bluetongue, *Tiliqua adelaidensis* (Squamata: Scincidae). Transactions of the Royal Society of South Australia. 1994;**118**(4):217-226

[9] Bull CM, Hutchinson MN. Saving the pygmy bluetongue lizard. In: Garnett S et al. editors. Recovering Australian Threatened Species: A Book of Hope. CSIRO Publishing; 2018. ProQuest Ebook Central. Available from: http://ebookcentral.proquest.com/lib/flinders/detail.action?docID=5311225

[10] The Commonwealth of Australia. *Tiliqua Adelaidensis* (Pygmy Blue-Tongue Lizard) Conservation Advice. Canberra: Department of Climate Change, Energy, the Environment and Water; 2023

[11] Dorrough J, Ash J, McIntyre S. Plant responses to livestock grazing frequency in an Australian temperate grassland. Ecography. 2004;**27**(6):798-810

[12] Hutchinson MF, McIntyre S, Hobbs RJ, Stein JL, Garnett S, Kinloch J. Integrating a global agro-climatic classification with bioregional boundaries in Australia. Global Ecology and Biogeography. 2005;**14**(3):197-212

[13] Williams M. The Making of the South Australian Landscape: A Study in the Historical Geography of Australia. New York: Academic Press; 1974. p. 518

[14] Moore RM. Ecological observations on plant communities grazed by sheep in Australia. In: Keast A, Crocker RL, Christian CS, editors. Biogeography and Ecology in Australia. Dordrecht, Netherlands: Springer; 1959. pp. 500-513

[15] Neagle N, editor. A Biological Survey of the Mid North and Yorke Peninsula, South Australia, 2003-2004: Assessment of Biodiversity Assets

at Risk. South Australia: Department for Environment and Heritage; 2008

[16] Jones PG. In: Angas GF, editor. Illustrating the Antipodes: George French Angas in Australia and New Zealand. Adelaide, SA: NLA Publishing; 2020. pp. 1844-1845

[17] Jessup RW. A vegetation and pasture survey of counties Eyre, Burra, and Kimber-ley, South Australia. Transactions of the Royal Society of South Australia. 1948;**72**:33-68

[18] Nicholls M. Native grasslands – A South Australian perspective. In: Barlow T, Thorburn R, editors. Balancing Conservation and Production in Grassy Landscapes. Clare, South Australia: Bushcare; 1999. p. 20

[19] Turner J. National Recovery Plan for the Iron-Grass Natural Temperate Grassland of South Australia Ecological Community 2012. South Australia: Department of Environment and Natural Resources; 2012

[20] Specht RL. The vegetation of South Australia. In: James AB, editor. 2nd ed. Canberra: Government Printer; 1972

[21] Hyde MK. The Temperate Grasslands of South Australia; their Composition and Conservation Status. Sydney: World Wide Fund for Nature Australia; 1995

[22] Axelrod DI. Rise of the grassland biome, Central North America. Botanical Review. 1985;**51**(2):163-201

[23] Foreman P. The Role of Aboriginal Burning in the Biogeography of the Temperate Grasslands of South-Eastern. Australia: La Trobe University; 2023

[24] Foreman PW. A framework for testing the influence of aboriginal burning on grassy ecosystems in lowland, Mesic South–Eastern Australia. Australian Journal of Botany. 2016;**64**(8):626-642

[25] Tindale NB. Tribal Boundaries in Aboriginal Australia. Canberra: Australian National University Press; 1974

[26] Graham J, Cullen B, Lodge G, Andrew M, Christy B, Holst P, et al. SGS animal production theme: Effect of grazing system on animal productivity and sustainability across southern Australia. Australian Journal of Experimental Agriculture. 2003;**43**(8):977-991

[27] South Australian Commercial Kangaroo Management Plan 2020-2024. Adelaide, SA, Australia: Government of South Australia, Department for Environment and Water; 2019

[28] Armstrong G, Reid J. The rediscovery of the Adelaide pygmy bluetongue *Tiliqua adelaidensis* (Peters, 1863). Herpetofauna. 1992;**22**(2):3-6

[29] Fenner A, Hutchinson M, McDonald P, Robertson P. *Tiliqua adelaidensis*. The IUCN Red List of Threatened Species 2018: e.T21902A101743579. 2018. DOI: 10.2305/IUCN.UK.2018-1.RLTS.T21902A101743579.en

[30] Cogger H. Reptiles and Amphibians of Australia. Ithaca, NY: Cornell University Press; 1992

[31] Fellows HL, Fenner AL, Bull CM. Spiders provide important resources for an endangered lizard. Journal of Zoology. 2009;**279**(2):156-163

[32] Clayton J. Impacts of Sheep Grazing on Burrow Use by Spiders and Pygmy Bluetongue Lizards (Tiliqua

Adelaidensis). [PhD thesis]. Adelaide, SA, Australia: Flinders University; 2018

[33] Milne T, Bull CM, Hutchinson MN. Use of burrows by the endangered pygmy blue-tongue lizard, *Tiliqua adelaidensis* (Scincidae). Wildlife Research. 2003;**30**(5):523-528

[34] Livieri TM, Forrest SC, Matchett MR, Breck SW. Conserving endangered black-footed ferrets: Biological threats, political challenges, and lessons learned. In: Imperiled: The Encyclopedia of Conservation. Lincoln, USA: Elsevier, University of Nebraska; 2022. pp. 458-470. DOI: 10.1016/B978-0-12-821139-7.00061-1

[35] Stevens TA, Evans MC, Osborne WS, Sarre SD. Home ranges of, and habitat use by, the grassland earless dragon (*Tympanocryptis pinguicolla*) in remnant native grasslands near Canberra. Australian Journal of Zoology. 2010;**58**(2):76-84

[36] Rix MG, Huey JA, Main BY, Waldock JM, Harrison SE, Comer S, et al. Where have all the spiders gone? The decline of a poorly known invertebrate fauna in the agricultural and arid zones of southern Australia. Austral Entomology. 2017;**56**(1):14-22

[37] Keller T, Colombi T, Ruiz S, Schymanski SJ, Weisskopf P, Koestel J, et al. Soil structure recovery following compaction: Short-term evolution of soil physical properties in a loamy soil. Soil Science Society of America Journal. 2021;**85**(4):1002-1020

[38] Bradshaw CJA. Little left to lose: Deforestation and forest degradation in Australia since European colonization. Journal of Plant Ecology. 2012;**5**(1):109-120

[39] Pettigrew M, Bull CM. The impact of heavy grazing on burrow choice in the pygmy bluetongue lizard, *Tiliqua adelaidensis*. Wildlife Research. 2011;**38**(4):299-306

[40] Souter NJ. Habitat Requirements and Conservation of the Endangered Pygmy Blue Tongue Lizard *Tiliqua Adelaidensis*. [PhD thesis]. Adelaide, SA, Australia: Flinders University of South Australia; 2003

[41] Souter NJ, Bull CM, Hutchinson MN. Adding burrows to enhance a population of the endangered pygmy blue tongue lizard, *Tiliqua adelaidensis*. Biological Conservation. 2004;**116**(3):403-408

[42] Nielsen TP, Bull CM. Effects of grazing regime on the relative body mass of the endangered pygmy bluetongue lizard (*Tiliqua adelaidensis*). Austral Ecology. 2019;**45**(4):460-467

[43] Ebrahimi M, Bull CM. Behavioural changes in an endangered grassland lizard resulting from simulated agricultural activities. Journal of Arid Environments. 2015;**113**:102-107

[44] Daniell TL, Hutchinson MN, Ainsley P, Gardner MG. Tussock grasses are an important resource for captive pygmy bluetongue lizards, *Tiliqua adelaidensis*. In: Daniell TL, editor. Behavioural Factors Affecting Translocation Success in the Pygmy Bluetongue Lizard. [PhD thesis]. Vol. 169. Adelaide, SA, Australia: Flinders University; 2021

[45] Schofield JA, Fenner AL, Pelgrim K, Bull CM. Male-biased movement in pygmy bluetongue lizards: Implications for conservation. Wildlife Research. 2012;**39**(8):677-684

[46] Schofield JA, Gardner MG, Fenner AL, Bull CM. Promiscuous mating in the endangered Australian lizard *Tiliqua adelaidensis*: A potential windfall

for its conservation. Conservation Genetics. 2014;**15**(1):177-185

[47] Clayton J, Bull CM. The impact of sheep grazing on burrows for pygmy bluetongue lizards and on burrow digging spiders. Journal of Zoology. 2015;**297**(1):44-53

[48] Ebrahimi M, Bull CM. Lycosid spiders are friends and enemies of the endangered pygmy bluetongue lizard (*Tiliqua adelaidenis*). Transactions of the Royal Society of South Australia. 2012;**136**:45-49

[49] Nielsen TP, Bull CM. Winter durability of pygmy bluetongue lizard burrows is higher for occupied than for unoccupied burrows and for those in less-grazed neighbourhoods. Wildlife Research. 2017;**43**(8):634-639

[50] Ebrahimi M, Schofield J, Bull C. Getting your feet wet: Responses of the endangered pygmy bluetongue lizard (*Tiliqua adelaidensis*) to rain induced burrow flooding. Herpetology. 2012;**5**:297-301

[51] Pettigrew M, Bull CM. The response of pygmy bluetongue lizards to simulated grazing in the field during three drought years. Wildlife Research. 2012;**39**(6):540-545

[52] Pettigrew M, Bull CM. Prey capture behaviour by pygmy bluetongue lizards with simulated grazing. New Zealand Journal of Ecology. 2014;**38**(1):45-52

[53] Nielsen TP, Fildes S, Bull CM. Does sheep grazing affect burrow choice of the pygmy bluetongue lizard (*Tiliqua adelaidensis*)? Journal of Arid Environments. 2017;**142**:22-30

[54] Michael KH, Baring R, Gardner MG. Is there a lizard down that spider burrow? Microhabitat influences spider burrow occupancy by the endangered pygmy bluetongue. Wildlife Research. DOI: 10.1071/WR23146

[55] Holechek JL, Gomez H, Molinar F, Galt D. Grazing studies: What we've learned. Rangelands Archives. 1999;**21**(2):12-16

[56] Mid North Grasslands Working Group Land Water and Wool. How to Make Money out of Grass – A Farmers Guide to Grazing Management of Native Pastures in the Northern Agricultural Districts of SA. Adelaide, SA, Australia: The Mid North Grasslands Working Group; 1986

[57] Morris CD. How biodiversity-friendly is regenerative grazing? Frontiers in Ecology and Evolution. 2021;**9**:816374

[58] Metera E, Sakowski T, Słoniewski K, Romanowicz B. Grazing as a tool to maintain biodiversity of grassland - A review. Animal Science Papers and Reports. (Jastrzębiec, Poland: Institute of Genetics and Animal Breeding). 2010;**28**(4):315-334

[59] Waters CM, Orgill SE, Melville GJ, Toole ID, Smith WJ. Management of grazing intensity in the semi-arid rangelands of southern Australia: Effects on soil and biodiversity. Land Degradation and Development. 2017;**28**(4):1363-1375

[60] Ebrahimi M, Godfrey SS, Fenner AL, Bull CM. Interactions between pygmy bluetongue lizards and co-existing species. Transactions of the Royal Society of South Australia. 2015;**139**(2):145-151

[61] Souter NJ, Bull CM, Lethbridge MR, Hutchinson MN. Habitat requirements of the endangered pygmy bluetongue

lizard, *Tiliqua adelaidensis*. Biodiversity and Conservation. 2007;**135**(1):33-45

[62] Alpe MJ, Kingery JL, Mosley JC. Effects of summer sheep grazing on browse nutritive quality in autumn and winter. The Journal of Wildlife Management. 1999;**63**:346-354

[63] Clayton J, Bull M. The impact of sheep grazing on the depth of spider burrows and of burrows selected by the pygmy bluetongue lizard (*Tiliqua adelaidensis*). Wildlife Research. 2017;**43**(8):691-703

[64] Smith AL, Gardner MG, Fenner AL, Bull CM. Restricted gene flow in the endangered pygmy bluetongue lizard (*Tiliqua adelaidensis*) in a fragmented agricultural landscape. Wildlife Research. 2009;**36**(6):466-478

[65] Johnston W. The place of C4 grasses in temperate pastures in Australia. New Zealand Journal of Agricultural Research. 1996;**39**(4):527-540

[66] Bull CM, Godfrey SS, Ebrahimi M, Fenner AL. Long and short term residence in refuge burrows by endangered pygmy bluetongue lizards. Amphibia-Reptilia. 2015;**36**(2):119-124

[67] Ebrahimi M, Godfrey SS, Fenner AL, Bull CM. Mating behaviour in pygmy bluetongue lizards: Do females 'attract' male lizards? Australian Journal of Zoology. 2014;**62**(6):491-497

[68] Delean S, Bull CM, Brook BW, Heard LMB, Fordham DA. Using plant distributions to predict the current and future range of a rare lizard. Diversity and Distributions. 2013;**19**(9):1125-1137

[69] Fordham DA, Watts MJ, Delean S, Brook BW, Heard LMB, Bull CM. Managed relocation as an adaptation strategy for mitigating climate change threats to the persistence of an endangered lizard. Global Change Biology. 2012;**18**(9):2743-2755

[70] Michael KH, Baring R, Gardner MG. Assessing the suitability of habitat for the endangered pygmy bluetongue. Endangered Species Research. DOI: 10.3354/esr01372

[71] Trewartha DM, Clayton JL, Godfrey SS, Gardner MG. Heat water and reptiles – Do the hydro-thermal properties of animals at the source location persist at the translocation site? Animal Conservation. 2024. DOI: 10.1111/acv.12942

[72] Welch JB, Venkateswaran A. The dual sustainability of wind energy. Renewable and Sustainable Energy Reviews. 2009;**13**(5):1121-1126

[73] Meat and livestock Australia. Biodiversity and Vegetation 2024. Available from: https://www.mla.com.au/research-and-development/Environment-sustainability/biodiversity-and-vegetation/ [Accessed: August 7, 2024]

[74] Chmelík V, Šarapatka B, Machač O, Mikula J, Laška V, Tuf IH. The effect of farming system and management practices on surface-dwelling soil macrofauna. Zemdirbyste-Agriculture. 2019;**106**:291-296

Chapter 4

Navigating the Dilemma of Human-Wildlife Conflict in the Selous-Niassa Ecosystem: Tanzania

Philipo J. Mtweve, Felician E. Chemihanda, Musa M. Bulim, Beevans B. Biseko and David Kaberege

Abstract

The Selous-Niassa Wildlife Corridor, spanning Tanzania and Mozambique, is a critical conservation area home to a diverse array of large mammal species. This study presents a comprehensive analysis of the dynamics within the corridor, with a focus on the escalating human-wildlife conflicts that threaten both local communities and the region's biodiversity. The findings indicate a rapid expansion of cropland, coupled with widespread deforestation and wildfires, which have significantly altered the landscape and exacerbated clashes between humans and wildlife. Crop raiding by large mammals, such as elephants and primates, has emerged as a dominant issue, resulting in substantial economic losses and food security challenges for local communities. Physical guarding remains the primary mitigation strategy, yet this approach is fraught with drawbacks, including heightened resentment toward conservation efforts and the risk of animal attacks. The time and resources invested by communities in protecting their crops detract from productive agricultural activities, further compounding the negative impacts. Addressing these complex issues will require a multifaceted approach. Enhancing community education and implementing effective compensation/consolation schemes could be crucial steps in fostering human-wildlife coexistence. Comprehensive studies to document the full range of species are also needed to gain a more holistic understanding of biodiversity dynamics.

Keywords: Selous-Niassa wildlife corridor, human-wildlife conflict, crop raiding, biodiversity conservation, community-based management

1. Introduction

The coexistence between humans and wildlife is crucial for preserving biodiversity and ensuring the well-being of both ecosystems and communities [1]. As human populations expand and encroach upon natural habitats, it is increasingly important to establish sustainable practices that allow for harmonious interaction between humans and wildlife [2].

In recent years, the rapid growth and redistribution of human populations has led to increased contact between people and wildlife, as villages expand into suburbs and suburbs expand into cities [3]. The expansion of agriculture and livestock farming near wilderness and protected areas has also accelerated, making human-wildlife coexistence challenging to achieve [4]. Human-wildlife conflict is a widespread issue, occurring in urban, rural, terrestrial, and aquatic realms, both in protected and unprotected areas that are shared by humans and animals [5]. The increasing prevalence of these shared spaces and their implications for biodiversity conservation and human well-being raises the important question: how can we respond to conflicts and achieve sustainable coexistence in these shared areas?

The Selous-Niassa Wildlife Corridor is no exception to human-wildlife conflict. This transboundary corridor links two major biodiversity ecosystems: The Selous Game Reserve (now Nyerere National Park), covering 50,000 km^2, and the Niassa Reserve, spanning over 42,000 km^2. The Selous Game Reserve, a UNESCO World Heritage Site, is one of Africa's largest protected areas and supports significant populations of elephants, buffalo, sable antelope, Lichtenstein's hartebeest, greater kudu, eland, Nyassa wildebeest, black rhinoceros, cheetah, giraffe, vervet monkeys, Sykes' monkeys, yellow baboons, bush pigs, hippopotamus, and crocodile, along with numerous migratory and resident bird species [6]. This makes it a critical habitat for in-situ conservation.

Similarly, the Niassa Reserve is Mozambique's largest protected area and is home to African wild dogs, sable antelope, elephants, over 400 bird species, and large populations of Cape buffalo, impala, wildebeest (including the Niassa wildebeest), Boehm's zebra, leopards, and Johnston's impala [7, 8]. Species recorded in both the Selous Game Reserve and Niassa Reserve have also been observed within the Selous-Niassa Wildlife Corridor, which spans 10,000 km^2 [6]. This indicates that the corridor facilitates the movement of migratory species between these two protected areas [9].

Nestled between Tanzania's Selous Game Reserve and Mozambique's Niassa National Reserve, the Selous-Niassa Wildlife Corridor serves as a vital conduit for biodiversity. Covering about 10,000 km2 and Stretching across 29 villages, where communities predominantly engage in subsistence farming and pastoralism, the corridor faces escalating human-wildlife conflict challenges exacerbated by climate change-induced environmental shifts. Lack of effective management of the corridor makes it vulnerable to human incursions, threatens the delicate balance of its ecosystem [6], and intensifies the human-wildlife corridor in the area.

Human-wildlife coexistence holds significant value in attaining both livelihood and biodiversity conservation in this region for multiple reasons. Firstly, wildlife plays a crucial role in maintaining healthy ecosystems by contributing to pollination, seed dispersal, and nutrient cycling, which are essential for the sustainability of plant communities and overall ecosystem health. Secondly, wildlife serves as a valuable economic asset, attracting tourists and generating revenue through ecotourism and wildlife-based activities. Lastly, the presence of wildlife enriches our lives, providing esthetic enjoyment and serving as a source of inspiration and wonder [10].

In the course of protecting the corridor, several initiatives have been put in place including the establishment of Wildlife management areas (WMAs). Mbarang'andu WMA is one of the WMAs formed in this important corridor. The corridor plays an important role in the management of the wildlife that migrates from either of the two protected areas (Selous or Niassa) [11]. The establishment of the corridor among other reasons was to manage the wildlife and enhance community livelihood through photographic and hunting tourism [11].

Although the corridor promotes conservation, community support has been challenging due to the area's remoteness and inadequate infrastructure, which have hindered the arrival of tourists [12]. While habitats have been restored, improving livelihoods remains challenging [12]. The ongoing recovery of the corridor's habitats has led to an increase in wildlife, particularly elephants, which has resulted in crop raiding and undermined communities' willingness to support conservation efforts [13]. The current trend of wildlife attacks on crops and people poses significant challenges for governments striving to align wildlife conservation with sustainable development. Despite being a pertinent corridor designated for protecting wildlife, wildlife often extends beyond the boundaries and utilizes areas outside these designated spaces [14].

Several studies have been conducted in the area, including research by Sengelela [13] and Noel and Kangalawe [12], which found that for two decades, the relationship between community participation in conservation and economic empowerment in Selous-Niassa, including the corridor, did not provide a useful solution for enhancing coexistence. Communities have continued to expand their farms and settlements closer to the corridor. The expansion of agriculture and human settlements in close proximity to the corridor has adversely affected the normal functioning of the ecosystem for several decades. Concurrently, increased efforts to control poaching in recent years have led elephants to roam outside the currently so-called a designated corridor and nearby protected areas, resulting in an increasingly significant issue of human-elephant conflict across all villages that are located in proximity to the corridor.

Due to the lack of alternative sources of revenue, communities that suffer crop losses to wildlife have been unable to capitalize on other opportunities and are increasingly resentful of wildlife, making them less likely to support wildlife management and conservation efforts. Studies conducted along the corridor indicate that, on average, 6.5 out of 10 households in these villages experience crop raiding annually. This loss translates to food insecurity and a strained relationship between management and livelihoods. Human-wildlife conflicts are escalating in all the 29 villages that comprise the corridor as wildlife crop raiders increase [13].

The corridor in the last 10 years was identified as a hotspot for wildlife poaching, particularly of elephants. Several interventions have been implemented to reduce poaching, but this has led to a rise in human-wildlife conflicts in the area. Non-governmental organizations (NGOs) and government institutions have been working to enhance coexistence between humans and wildlife in the corridor. However, despite these efforts, cases of human-wildlife conflict within the corridor have remained high [15]. The ongoing recovery of wildlife populations, especially elephants, has contributed to increased crop raiding and undermined community support for conservation efforts in the region.

This study therefore aimed to:

1. Identify the key drivers of human-wildlife conflicts within the Selous-Niassa Wildlife Corridor.

2. Determine the primary types of human-wildlife conflicts present in the Selous-Niassa Wildlife Corridor.

3. Identify the animal species most commonly involved in these conflicts.

4. Evaluate the strategies that have been implemented to address and mitigate these human-wildlife conflicts.

2. Methods

2.1 Study area description

The study was conducted in southern Tanzania's Selous-Niassa Wildlife Corridor (SNWC), **Figure 1**, and confined itself to seven villages, namely: Kitanda, Nambecha, Likuyu Sekamaganga, Kilimasera, Songambele, Mchomoro, and Mtelawamwahi. This vital wildlife corridor bridges East Africa's two most significant protected areas: The Selous Game Reserve (now divided into Selous Game Reserve and Nyerere National Park) in Tanzania and the Niassa Reserve in Mozambique. Together, they form a vast conservation ecosystem spanning roughly 154,000 square kilometers [6]. The SNWC is renowned for being one of the world's largest elephant ranges, with an estimated population of 85,000 [16]. It is also a haven for vast buffalo herds, wild dog packs, migratory birds, and lions [17]. This region is crucial for conserving migratory species, including the remaining 13% of the world's wild dog population, savannah elephants, leopards, buffalos, and others that traverse between the three protected areas [6]. An estimated 15,000 elephants migrate between the Niassa Reserve and Selous Game Reserve, driven by seasonal changes in forage availability. Due to its rich biodiversity, this area is critical to wildlife conservation efforts. Consequently, various conservation organizations recognize this corridor as a key hub for wildlife movement [12].

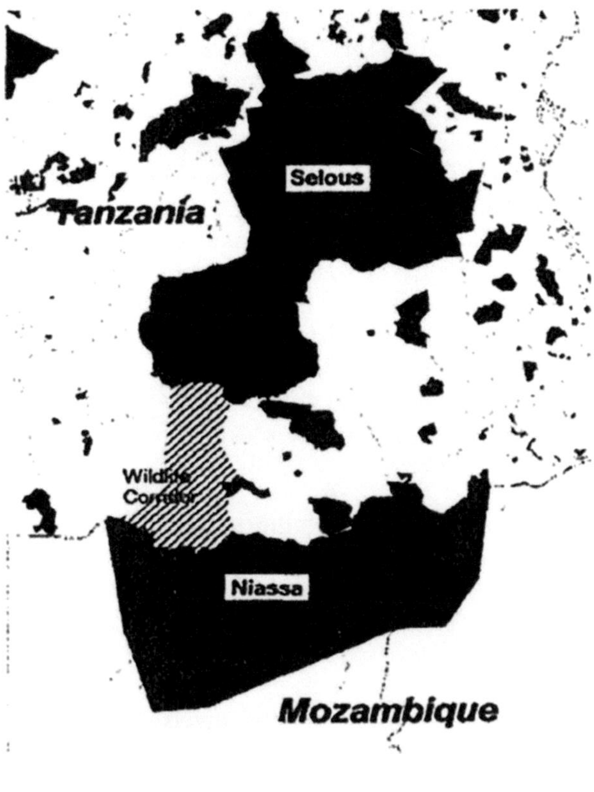

Figure 1.
The selous-Niassa wildlife corridor.

2.2 Data collection

2.2.1 Cropland distribution

To investigate the growing human-wildlife conflict in the area, we analyzed cropland distribution and expansion within the corridor and a 2.5 km buffer around the wildlife corridors. Cropland data was obtained from the Global Cropland Extent raster developed by the Global Land Analysis and Discovery Lab, with a 30 m spatial resolution [18]. The rasters were downloaded, projected to EPSG:32736 - WGS 84 /UTM zone 36S using QGIS, and then subjected to zonal statistics to extract the cropland values in acres for the years 2003, 2007, 2011, 2015, and 2019 within the corridor and the 2.5 km buffer.

2.2.2 Wildfire burning

To assess the annual burned area in the corridor, we extracted yearly burned data from the MOD44W.006 datasets [19]. We obtained data from 2001 to 2022, which has a spatial resolution of 250 meters [20].

We utilized the Google Earth Engine (GEE) environment to extract and process the data. The data was projected into the "EPSG:32736 - WGS 84/UTM zone 36S" coordinate system to ensure consistency.

2.2.3 Semi-structured household interviews

To assess human-wildlife conflict and the extent of depredations, we conducted semi-structured interviews with households. Our team carried out one-on-one interviews to gather comprehensive insights into the coexistence of humans and wildlife within the corridor. The sampling unit for this study was the household. We distributed questionnaires to 5% of the households in each of the 7 villages included in the study. A random sampling approach was used to select the households, with random sampling points generated in QGIS [21]. Prior to the main survey, we conducted a pretest of the questionnaire with 30 households. The data from these pretest households was excluded from the final analysis. In total, we administered 405 questionnaires across the 7 villages.

2.2.4 Key informant interviews

In addition to the household interviews, we conducted key informant interviews with individuals who hold pivotal roles in working with the local communities. This group of key informants consisted of two district game officers, three members of the Mbarang'andu Wildlife Management Area (WMA) management team, 14 environmental committee members, and 5 village executive officers. The purpose of these key informant interviews was to gather essential background information about the corridor, gain a deeper understanding of the challenges it faces, and collect valuable insights regarding Human-Wildlife Conflict in the area.

2.2.5 Focus group discussion

To further explore the various perspectives within the local communities, we organized focus group discussions. Each of the 7 villages hosted one such discussion, with groups ranging in size from 6 to 8 individuals representing diverse backgrounds and areas of expertise. During these focus group sessions, it became evident how challenging it can be to reach a consensus on issues related to human-wildlife coexistence. This realization highlighted the importance of initiating environmental education efforts within the communities. The goal of such education would be to enable the development of a shared understanding of the value of wildlife, strategies for fostering harmonious coexistence between humans and wildlife, and a common framework that can accommodate the diverse interests within the corridor.

Each focus group discussion lasted between 45 and 60 minutes, and we employed probing, follow-up, and exit questions to facilitate meaningful conversations. Additionally, we used a smartphone for audio recording and notebooks to document the comments made by participants.

3. Results

3.1 Evaluating Forest loss and fires as indicators of increasing human-wildlife conflicts within the corridor

Our central hypothesis was that increases in cropland areas and burned landscapes within the corridor would be associated with a higher likelihood of human-wildlife conflicts. The results obtained supported this assumption, indicating that the study area has experienced an overwhelming expansion of cropland over the last 16 years. The area's cropland has been expanding at an accelerating rate, with percent increases of 1.52% from 2003 to 2007, 52.31% from 2007 to 2011, 99.82% from 2011 to 2015, and 52.14% from 2015 to 2019 (**Figure 2**). Overall, the general percent increase in cropland from 2003 to 2019 was 370%, highlighting the significant and intensifying trend of cropland expansion within the corridor over this 16-year period.

In addition to the cropland dynamics, our analysis also examined patterns of burned area and forest loss within the corridor. The data indicated that, on average, 550 km^2 have been burned annually over the last 23 years, while approximately 5 km^2 have been deforested each year for the past 22 years (**Figure 2**). These findings further underscore the environmental changes occurring in the region, which may be exacerbating human-wildlife conflicts.

3.2 What are the primary types of human-wildlife conflicts recorded within the Selous-Niassa wildlife corridor?

Crop raiding and destruction have been reported as a significant and pervasive issue within the corridor. According to the interviews conducted, a substantial majority of 398 out of 405 respondents identified crop raiding as a major problem. This widespread concern underscores the gravity of the issue, indicating that local communities face considerable challenges related to wildlife entering agricultural areas and causing damage, potentially leading to economic losses and food security concerns.

Infrastructure destruction was also recognized as a problem by 102 respondents. While not as prevalent as crop raiding, it remains a substantial concern, indicating the tangible impact of human-wildlife conflicts on local infrastructure and property. On the other hand, encroachment emerged as a relatively lesser-reported issue among the conflicts (**Figure 3**). However, the analysis of cropland datasets suggests that communities are expanding toward the corridor, accelerating crop raiding and posing a challenge to human-wildlife coexistence efforts.

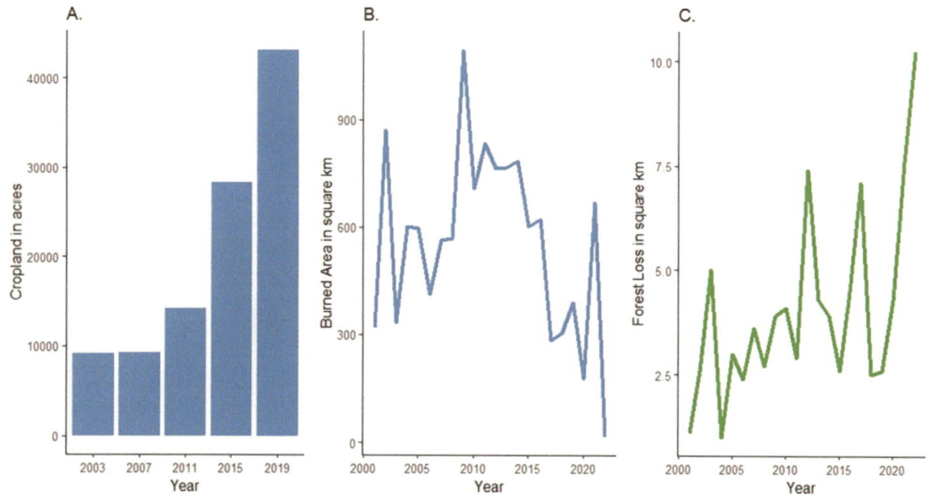

Figure 2.
(A) the cropland expansion for the last 16 years, (B) yearly burned areas over the last 22 years, and (C) deforestation over the last 22 years. All these data were extracted within and between the 2.5 km buffer within the Selous-Niassa Wildlife Corridor.

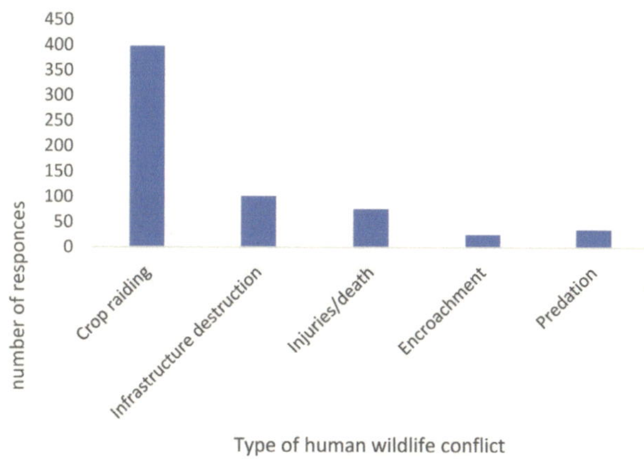

Figure 3.
Reported types of human-wildlife conflict in the Selous-Niassa Wildlife Corridor.

3.3 Which animals are involved in human-wildlife conflicts within the corridor?

The study revealed a diverse range of species involved in human-wildlife conflicts, as reported by local communities. The savannah elephant emerged as the most prominent and destructive species, garnering significant attention due to its interactions with human settlements and agricultural activities (**Figure 4**). The presence of these large and powerful animals in close proximity to communities posed substantial challenges, including crop destruction and safety concerns. This also indicates how ongoing deforestation, cropland expansions, and wildfires have influenced the distribution of elephants within community lands.

Following closely behind the savannah elephant, yellow baboons also featured prominently in human-wildlife conflicts. These primates, known for their opportunistic foraging habits, were reported to affect local communities by raiding crops and

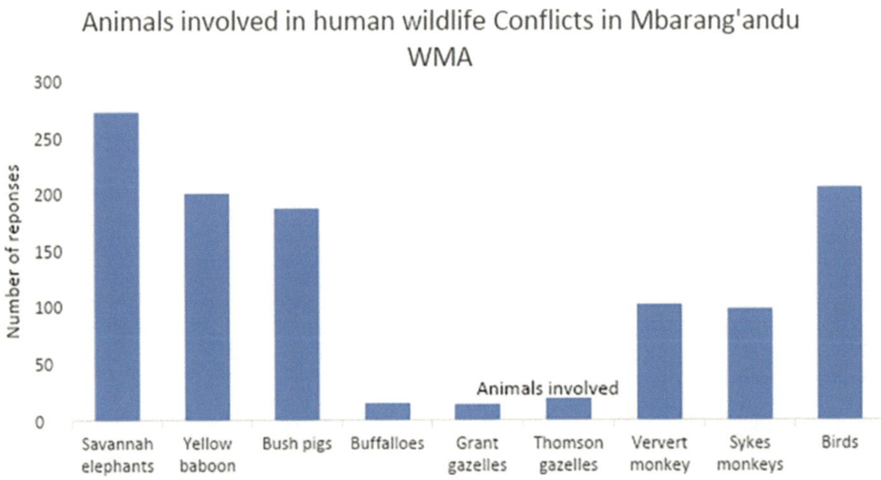

Figure 4.
The prominent wild animals reported to cause human-wildlife conflict within the Selous-Niassa wildlife corridor.

sometimes causing damage to property. Furthermore, the study highlighted the impact of birds on local communities, specifically in the context of rice/paddy and sunflower farming. The species involved in crop raiding were commonly referred to as the red-billed quelea, doves, francolins, and pied crows. Additionally, communities are experiencing intensive poultry loss from birds of prey, including kites, pied crows, and eagles.

3.4 What activities influence the human-wildlife conflict within the corridor?

During the interviews and focus group discussions, communities affirmed that crop farming was a primary driver of human-wildlife conflict. A significant 342 respondents mentioned that the presence of farms attracts wildlife, particularly crop raiders and birds. This perception aligns with the data from the Global Cropland Extent raster, which indicated that over the last 16 years, the corridor has witnessed a rapid expansion of farmlands, almost a 370% increase. This reflects the increasing competition for land use and the communities' proximity to the corridor.

Furthermore, the mismanagement of livestock was recognized by 102 respondents as a factor contributing to conflicts. This suggests that the way in which livestock are managed increases their vulnerability to predation by wildlife and birds of prey. Livestock are often left to free-range in the village lands, where they can be easily accessed by birds of prey and small carnivores. Interestingly, 63 communities indicated that they were unsure about the reasons for wildlife invading their farms or preying on their livestock (**Figure 5**). This lack of clarity could signify a need for increased awareness and education within these communities to better understand and address the complexities of human-wildlife interactions.

3.5 What strategies have been implemented to address and mitigate these conflicts?

The study explored the strategies communities have put in place to deal with problem animals. The most prevalent strategy, reported by 289 respondents, was

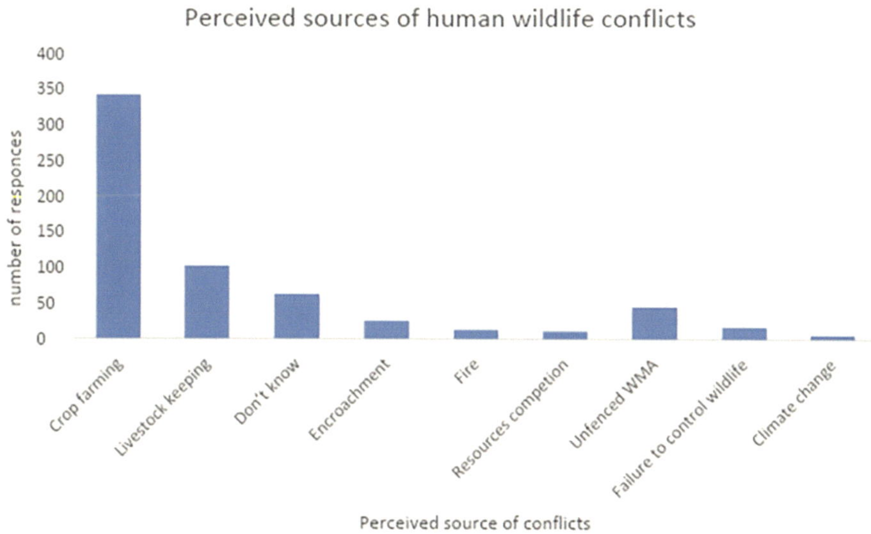

Figure 5.
The communities perceived human-wildlife conflict sources in the study area.

physical guarding. In this approach, communities spend more than 10 hours daily on their farms guarding their crops. However, communities seemed very irritated with this work, as it was more time-consuming than the production itself. This approach was also associated with increased incidences of human injuries and death. Communities felt betrayed by all conservation organizations and, in particular, the government.

Physical guarding was followed by snaring and hunting problem animals, which was the second-mentioned strategy (102). This act puts the corridor's survival in question, as it may bring unintended ecological impacts. Patrols conducted by rangers were mentioned by 57 respondents, but communities complained that the response took too long for rangers to appear. Scarecrows, reported by 43 respondents, were another strategy mentioned, but their effectiveness varied depending on the species involved and their familiarity with such deterrents. Surprisingly, 56 respondents indicated that they employed no specific strategy to combat human-wildlife conflicts, suggesting a lack of proactive measures or a reliance on traditional coping mechanisms that may not effectively address the challenges posed by wildlife (**Figure 6**).

3.6 Perceived strategies for the long-term human-wildlife conflicts mitigations

The survey explored the strategies that communities perceive as effective in reducing human-wildlife conflict and fostering long-term coexistence within the corridor.

The most frequently mentioned strategy was community education and training, reported by 309 respondents. This approach involves educating community members about wildlife behavior, conflict prevention techniques, and coexistence strategies. By promoting understanding and reducing conflicts, community education and training

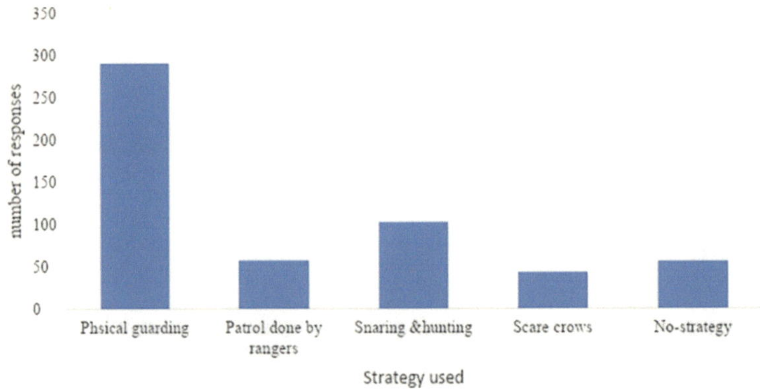

Figure 6.
Strategies used by communities on reducing depredations from wildlife.

will play a crucial role in establishing harmonious relationships between humans and wildlife. Another important strategy, mentioned by 300 respondents, was compensation. This seeks to alleviate the economic burden caused by wildlife conflicts and encourage tolerance toward wildlife by providing financial or material support to communities.

Increased ranger presence, reported by 197 respondents, was another highlighted strategy, requiring the deployment of more rangers to areas experiencing human-wildlife conflicts. This will enhance conflict management efforts, including monitoring wildlife activities, responding to conflicts, and enforcing regulations. Additionally, 167 respondents emphasized the importance of cooperation between conservation organizations and communities, emphasizing the need for collaborative partnerships and shared responsibility in addressing human-wildlife conflicts. By working together, more effective and sustainable solutions can be developed and implemented.

The survey also explored the role communities believe they could play in managing human-wildlife conflicts. The majority of respondents, 390 communities, believe that reporting incidents to the government and having the government respond actively could solve the problem. This includes reporting conflicts to relevant authorities like wildlife departments or conservation agencies, and the responsible institutions providing compensation.

However, some communities held different perspectives. 208 communities believe that burning all the forests around the villages would prevent wildlife from being attracted to the area. 135 communities felt that there is no way to sustainably maintain both communities and wildlife in the region. 102 respondents believe that wildlife shooting and land use planning could help resolve the conflicts. Furthermore, 98 communities responded that clearing forests to create a buffer between human settlements and wildlife habitats would be an important strategy (**Figure** 7).

These divergent views highlight the complex and multifaceted nature of human-wildlife conflicts, where communities are seeking a variety of solutions, some of which may have significant ecological implications. Addressing these conflicts will require carefully balancing the needs and perspectives of both communities and wildlife conservation (**Figure** 7).

DOI: http://dx.doi.org/10.5772/intechopen.1006653

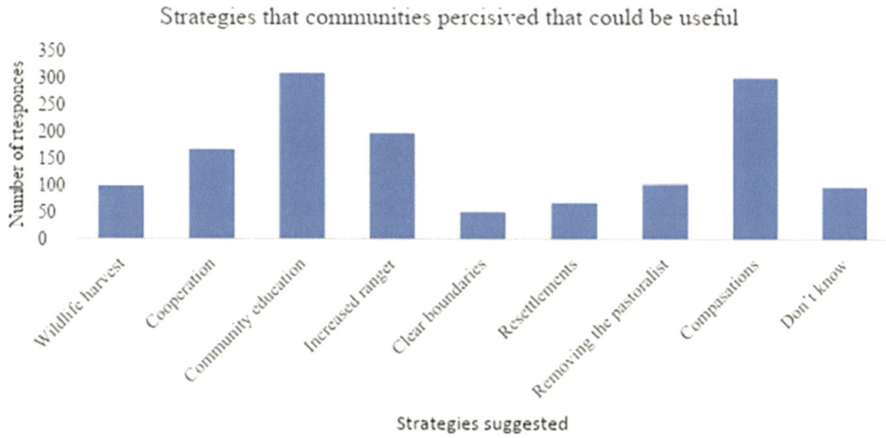

Figure 7.
Communities' perception toward participation in the management of resources within the corridor.

4. Discussion

Through the analysis, we observed several patterns of cropland expansion, settlement distributions, and wildfire, which we hypothesized could be one of the reasons why the area is experiencing intensive human-wildlife conflict. The central hypothesis of our study was that increases in cropland area, deforestation, and burned landscapes within the corridor would be associated with a higher likelihood of human-wildlife conflicts.

The results of the study supported this assumption. The data showed that the study area has experienced a significant and accelerating expansion of cropland over the last 16 years. Cropland area increased by 370% from 2003 to 2019, highlighting the intensifying trend of cropland expansion within the corridor. In addition to the cropland dynamics, the analysis also examined patterns of burned area and forest loss within the corridor. The data indicated that, on average, 550 km^2 have been burned annually over the last 23 years, while approximately 5 km^2 have been deforested each year for the past 22 years. These findings further underscore the environmental changes occurring in the region, which may be exacerbating human-wildlife conflicts.

The implications of these findings are significant for the management and conservation of the Selous-Niassa Wildlife Corridor. The rapid expansion of cropland and the associated deforestation and wildfire patterns suggest that the natural habitats and corridors for wildlife are under increasing pressure [22, 23]. This can lead to a loss of biodiversity, fragmentation of wildlife populations, and more frequent encounters between humans and wildlife, potentially resulting in increased conflicts and negative impacts on both communities and wildlife [24].

The analysis of the primary types of human-wildlife conflicts recorded within the Selous-Niassa Wildlife Corridor provided critical insights into the nature and scale of these interactions.

Crop raiding and destruction emerged as a significant and pervasive issue within the corridor. According to the interviews conducted, a substantial majority of 398 out of 405 respondents identified crop raiding as a major problem. This widespread concern underscores the gravity of the issue, indicating that local communities face considerable challenges related to wildlife entering agricultural areas and causing damage, potentially leading to economic losses and food security concerns.

Infrastructure destruction was also recognized as a problem by 102 respondents. Although not as prevalent as crop raiding, it remains a substantial concern, highlighting the tangible impact of human-wildlife conflicts on local infrastructure and property. In contrast, encroachment emerged as a relatively lesser-reported issue among the conflicts. However, the analysis of cropland datasets suggests that communities are expanding toward the corridor, accelerating crop raiding and posing a challenge to human-wildlife coexistence efforts.

The dominance of crop raiding as a primary issue underscores the need for targeted interventions and strategies to mitigate the negative impacts on local livelihoods and food security [24]. Addressing infrastructure destruction and encroachment issues can also contribute to a more holistic approach to managing these conflicts and promoting sustainable coexistence between humans and wildlife. The escalating incidence of crop raiding in protected areas is closely linked to activities at their peripheries, encompassing the cultivation of crops preferred by wildlife, encroachment intensifying the isolation of these areas [25], and ongoing rainfall failures pushing both communities and wildlife closer to water sources for agriculture [26].

The intrusion of wildlife into community farms has resulted in significant costs, with Savannah elephants emerging as major culprits. Although their visits may not be frequent, the impact is substantial, often leading to human injuries and even fatalities, a finding consistent with the observations of [27]. Another challenge arises from primates, which demonstrate a remarkable ability to adapt to strategies devised to deter them from raiding farms. Their diverse nature complicates matters further, with velvet and Sykes monkeys causing trouble during the early stages of crop production, while yellow baboons become culprits as the crops mature. Bush pigs also pose a significant challenge, as reported by numerous respondents. Their ability to infiltrate farms during both day and night adds complexity to efforts aimed at managing and controlling them. Additionally, birds prove to be particularly destructive in sunflower and rice farms. Interestingly, these avian threats are not frequently acknowledged by communities, possibly due to their exclusion from conservation discussions compared to more prominently mentioned rodents. However, this oversight may be problematic, as communities often use snares, inadvertently risking the loss of more species in the area.

Reports suggest that Tanzania is experiencing an unprecedented rate of species loss, potentially attributed to such overlooked factors [28]. Notably, no comprehensive study has documented the bird species or rodentia species in the area, highlighting a significant gap in our understanding of the biodiversity dynamics in these regions. Generally, Crop raiding impact by multiple animals worsens the problem which necessitates communities to invest considerable time and effort in safeguarding their farms.

To unravel the intricacies of human-wildlife conflicts in the area, an examination of community activities revealed that 99% of the respondents were engaged in either crop farming or livestock keeping. These activities are associated with the need for habitat which is identified as a primary driver of human-wildlife conflict [2, 29]. The complexity of the findings becomes evident as a majority of the communities expressed a preference for establishing new farms. Notably, these new farms are situated in close proximity to the Wildlife corridor. This choice is driven by the understanding that uncultivated land near the corridor tends to yield good harvests with minimal agricultural inputs, such as fertilizers and pesticides.

Of greater concern is the lack of awareness among communities regarding the corridor potential in the management of the wildlife to the two proximity protected areas that include Selous and Niassa Reserves. This implies that the well-being of animal populations in the corridor is intricately linked to the forage status in either the Niassa Reserve or the Selous Game Reserve. Consequently, the area consistently hosts wildlife in transit, making their movements challenging to manage and predict. This lack of awareness about the dynamic nature of the corridor and its connection to larger reserves adds an additional layer of complexity to the human-wildlife conflict scenario.

Our investigation also delved into the strategies implemented to foster human-wildlife coexistence or mitigate the adverse impacts caused by wildlife. The predominant approach reported by the majority was physical guarding. However, this strategy has significant drawbacks, contributing to heightened resentment within communities toward wildlife conservation. The necessity of spending considerable time safeguarding crops detracts from actual production activities, intensifying negative sentiments. Moreover, physical guarding is closely linked to the risk of animal attacks, exacerbating the challenges faced by communities.

This firsthand experience is vividly expressed by community members, who noted that over the past four years, 15 individuals have either been injured or killed by elephants or buffalos. This alarming trend resonates with similar occurrences documented in other parts of Africa. For instance, the Food and Agriculture Organization reported that in Mozambique's Cabo Delgado province, lions claimed the lives of 70 individuals over an 18-month period between 2001 and 2002. Notably, many of these victims were attempting to protect their crops from elephants, underscoring the grave consequences of the strategies employed to mitigate human-wildlife conflicts. Snaring and hunting emerged as the second most used strategy. This strategy raised the question of species loss in the areas.

Given the complexity of the problem and the strategies proposed so far, we also intended to find out what could be the way forward toward enhancing human-wildlife conflict (**Figure 8**). Respondents were optimistic that if community education and compensation/consolation could be made effective, the majority could live in harmony with wildlife and could be happy to share and work in collaboration with conservation organizations. A well-informed public may be the most important way to preserve native species and decrease instances of human-wildlife conflict. Similarly, according to [30], compensation schemes are good tools to distribute the costs between those who benefit from conservation and those who suffer the cost of damages and hope to reduce the consequences of human-wildlife conflicts.

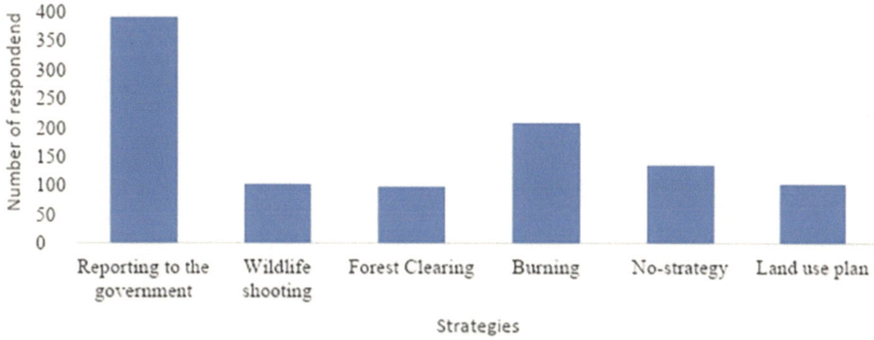

Figure 8.
How communities perceive their role in enhancing human-wildlife coexistence.

5. Conclusions

The comprehensive analysis of the Selous-Niassa Wildlife Corridor has revealed critical insights into the complex dynamics of human-wildlife interactions in the region. The findings highlight the rapid and accelerating expansion of cropland, coupled with significant patterns of deforestation and wildfires, which appear to be exacerbating the incidence of human-wildlife conflicts. The dominance of crop raiding as a primary issue, particularly by large mammals such as elephants and primates, underscores the severe economic and food security challenges faced by local communities.

To address these complex issues, a multifaceted approach is necessary. Enhancing community education and implementing effective compensation/consolation schemes could be crucial steps in fostering human-wildlife coexistence. Additionally, comprehensive studies to document the full range of species, including birds and rodents, are needed to gain a more holistic understanding of the biodiversity dynamics in the region. By addressing the underlying drivers of human-wildlife conflicts and promoting collaborative approaches that balance the needs of both communities and wildlife, the conservation and management of the Selous-Niassa Wildlife Corridor can be significantly improved.

Acknowledgements

We are grateful for the generous support of the Bio-Bridge Initiative (BBI) and International Elephant Fund (IEF), who have been pivotal in providing the necessary funding. Additionally, we acknowledge the ongoing assistance and collaboration of the Natural Resources Department, the Tanzania Commission for Science and Technology, and the Tanzania Wildlife Research Institute. Their efforts in connecting the project to local communities and facilitating the required permits have been invaluable.

Conflict of interest

The authors declare no conflict of interest.

Author details

Philipo J. Mtweve*, Felician E. Chemihanda, Musa M. Bulim, Beevans B. Biseko and David Kaberege
Environmental Conservation for Wildlife and Community Enterprise (ECOWICE), Morogoro, United Republic of Tanzania

*Address all correspondence to: philipo.mtweve@sua.ac.tz

References

[1] Carter NH, Linnell JDC. Building a resilient coexistence with wildlife in a more crowded world. Yortsos Y, editor. PNAS Nexus. 2023,**2**(3):pgad030

[2] Mekonen S. Coexistence between human and wildlife: The nature, causes and mitigations of human wildlife conflict around Bale Mountains National Park, Southeast Ethiopia. BMC Ecology. 2020;**20**(1):51

[3] Estes AB, Kuemmerle T, Kushnir H, Radeloff VC, Shugart HH. Land-cover change and human population trends in the greater Serengeti ecosystem from 1984-2003. Biological Conservation. 2012;**147**(1):255-263

[4] Barbosa JM, Hiraldo F, Romero MÁ, Tella JL. When does agriculture enter into conflict with wildlife? A global assessment of parrot–agriculture conflicts and their conservation effects. Diversity and Distributions. 2021;**27**(1):4-17

[5] Horgan FG, Kudavidanage EP. Farming on the edge: Farmer training to mitigate human-wildlife conflict at an agricultural frontier in South Sri Lanka. Crop Protection. 2020;**127**:104981

[6] Baldus RD, Hahn R. The Selous - Niassa Wildlife Corridor in Tanzania: Biodiversity Conservation from the Grassroots: Practical Experiences and Lessons from Integrating Local Communities into Trans-Boundary Natural Resources Management. Budakeszi, Rome: CIC Administrative Office. Food and Agriculture Organization; 2009

[7] Allan J, Grossmann F, Craig R, Nelson A, Maina J, Flower K, et al. Patterns of forest loss in one of Africa's last remaining wilderness areas: Niassa National Reserve (northern Mozambique). Parks. 2017;**23**(2):39-50

[8] Mittermeier RA, Mittermeier CG, Brooks TM, Pilgrim JD, Konstant WR, Da Fonseca GAB, et al. Wilderness and biodiversity conservation. Proceedings of the National Academy of Sciences. 2003;**100**(18):10309-10313

[9] Eakin E. The Effectiveness of Wildlife Corridors in Biological Conservation, a Case Study of the African Elephant (Loxodonta africana) in Tanzania. Plymouth, United Kingdom: Plymouth University; 2017. Available from: http://rgdoi.net/10.13140/RG.2.2.24238.36161 [Accessed: May 27, 2024]

[10] Gross E, Jayasinghe N, Brooks A, Polet G, Wadhwa R, Hilderink-Koopmans F. A Future for all: The Need for Human-Wildlife Coexistence. Gland, Switzerland: WWF; 2021

[11] Tarimo E. Resource Zone Management Plan (2009-2013) for the Mbarang'andu Pilot Wildlife Management Area. Dodoma, Tanzania: Tanzania Wildlife management Authority; 2009

[12] Noe C, Kangalawe RY. Wildlife protection, community participation in conservation, and (dis) empowerment in southern Tanzania. Conservation and Society. 2015;**13**(3):244

[13] Sengelela M. Local People Perceived Benefits and Costs of Trans-Frontier Conservation Areas: The Case of Likuyusekamaganga Village in Selous-Niassa Trans-Frontier Conservation Area. Netherlands: Wageningen University; 2013

[14] Halladay P. Conserving Biodiversity outside Protected Areas: The Role of

Traditional Agro-Ecosystems. United States: University of Minnesota: IUCN--the World Conservation Union; 1995

[15] Picard C. The Promise and Peril of Large-Scale Conservation: An Appraisal of the Selous Niassa Wildlife Corridor. Tanzania: Yale University; 2010

[16] Hofer H, Hildebrandt TB, Göritz F, East ML, Mpanduji DG, Hahn R, et al. Distribution and Movements of Elephants and Other Wildlife in the Selous-Niassa Wildlife Corridor. Tanzania. Eschborn, Germany: Tropical Ecology Support Programme, GTZ; 2004

[17] Mpanduji DG, East M, Hofer H. Analysis of habitat use by and preference of elephants in the Selous-Niassa wildlife corridor, southern Tanzania. African Journal of Ecology. 2009;**47**(2):257-260

[18] Potapov P, Turubanova S, Hansen MC, Tyukavina A, Zalles V, Khan A, et al. Global maps of cropland extent and change show accelerated cropland expansion in the twenty-first century. Nature Food. 2021;**3**(1):19-28

[19] Carroll M, DiMiceli C, Wooten M, Hubbard A, Sohlberg R, Townshend J. MOD44W MODIS/Terra Land Water Mask Derived from MODIS and SRTM L3 Global 250m SIN Grid V006 [Internet]. NASA EOSDIS Land Processes Distributed Active Archive Center; 2017. Available from: https://lpdaac.usgs.gov/products/mod44wv006/ [Accessed: Jun 14, 2024]

[20] Giglio L, Schroeder W, Justice CO. The collection 6 MODIS active fire detection algorithm and fire products. Remote Sensing of Environment. 2016;**178**:31-41

[21] QGIS.org. QGIS Geographic Information System. Open Source Geospatial Foundation Project; 2023. Available from: http://qgis.org

[22] Abade L, Macdonald DW, Dickman AJ. Using landscape and bioclimatic features to predict the distribution of lions, leopards and spotted Hyaenas in Tanzania's Ruaha landscape. Margalida a, editor. PLoS One. 2014;**9**(5):e96261

[23] Dolrenry S, Stenglein J, Hazzah L, Lutz RS, Frank L. A Metapopulation approach to African lion (Panthera leo) conservation. Roca AL, editor. PLoS One. 2014;**9**(2):e88081

[24] Barua M, Bhagwat SA, Jadhav S. The hidden dimensions of human–wildlife conflict: Health impacts, opportunity and transaction costs. Biological Conservation. 2013;**157**:309-316

[25] Walelign SZ, Nielsen MR, Jacobsen JB. Roads and livelihood activity choices in the greater Serengeti ecosystem, Tanzania. Muboko N, editor. PLoS One. 2019;**14**(3):e0213089

[26] Abram NK, Lim HY, Ng CKC, Yorath S, Razi MHM, Ong CGS, et al. The socio-economic and cultural impacts of the pan Borneo highway on indigenous and local communities in Sabah, Malaysian Borneo. Liew TS, editor. PLoS One. 2022;**17**(6):e0269890

[27] Osborn FV, Parker GE. Linking two elephant refuges with a corridor in the communal lands of Zimbabwe. African Journal of Ecology. 2003;**41**(1):68-74

[28] Mtui D, Owen-Smith N, Lepczyk C. Assessment of wildlife populations trends in three protected areas in Tanzania from 1991 to 2012. African Journal of Ecology. 2017;**55**(3):305-315

[29] Frank B. Human–wildlife conflicts and the need to include tolerance and coexistence: An introductory comment. Society and Natural Resources. 2016;**29**(6):738-743

[30] Upadhyay S. Challenges of Compensation Schemes for Human-Wildlife Conflict Mitigation. 2013. Available from: http://rgdoi.net/10.13140/2.1.2680.7680 [Accessed: June 14, 2024]

Chapter 5

Facing the Red Palm Weevil (*Rhynchophorus ferrugineus*) (Olivier) (Coleoptera: Curculionidae) – Safeguarding *Phoenix canariensis* and beyond

Meriem Msaad Guerfali, Kamel Charaabi, Mohamed Chaouachi, Sana Bouzembila, Salma Fadhl, Mariam Boudabouz and Haythem Hamden

Abstract

Phytosanitary regulations in various countries govern the importation and movement of plants and plant products. These regulations aim to prevent the introduction of harmful organisms into the territory, which may be present on plants and in the soil and can adversely affect both crops and biodiversity. The introduction of plant varieties carrying pests can have disastrous consequences. One significant example is the Red Palm Weevil (*Rhynchophorus ferrugineus*), inadvertently introduced to Tunisia with *Phoenix canariensis*. First detected in Carthage, the pest has since spread to several governorates, including Bizerte, Ariana, Ben Arous, and Zaghouan. The Red Palm Weevil, with a wide host range and distribution, threatens palm biodiversity by infesting palm trees, leading to their death. In Tunisia, this invasion poses a major risk, particularly as the *Phoenix dactylifera*, one of the preferred species of the RPW, is mainly found in the southern region of the country. Tunisia's status as the leading exporter of dates to the European Union further underscores the economic and agricultural implications of RPW infestation. To address this threat, various management strategies have been implemented worldwide and in Tunisia.

Keywords: *Rhynchoforus*, *P. canariensis*, *P. dactylifera*, biodiversity, conservation

1. Introduction

Biodiversity is essential to life itself, because it forms the interconnection of life on which we rely for. Biodiversity is important for our well-being and the survival of our species because it provides essential elements like food. Therefore, it is broadly acknowledged that invasive species pose a significant risk to biodiversity. Pests are one of the main drivers of biodiversity erosion; they represent a real threat, especially in the current context of climate change, and the measures taken to combat them can have unintended social, economic, and environmental repercussions.

The red palm weevil (RPW), *Rhynchophorus ferrugineus* (Olivier), is a coleoptera that consumes the palm trunks. This behavior destroys the meristem, which eventually kills the palm tree after high infestation degree. Originating in Southeast Asia, it spread to the Arabian Peninsula *via* infested material in the 1980s [1], crossing the Red Sea to North Africa in the 1990s to reach countries such as Tunisia and Morocco and then Europe, including Italy, France, and Spain [2]. The RPW feeds on a variety of palms from its natural habitat, including coconuts (*Cocos nucifera*), date palms (*Phoenix dactylifera*), sago palm (*Metroxylon sagu*), and oil palms (*Elaeis guineensis*) [3].

In the Mediterranean region, it has become the major pest of palms, mainly *Phoenix canariensis* Hort. ex Chabaud [4]. *P. canariensis*, which is of considerable horticultural interest, has been officially recognized as naturalized in various parts of the world. This includes Europe (specifically Spain, France, and Italy), northwestern Africa (Morocco and Tunisia), Mexico, South America, and the southern USA (California and Florida) (**Figure 1**) [5]. It is one of the most valued ornamental trees in subtropical and tropical regions due to its adaptability to a wide variety of locations [6]. *Phoenix canariensis* is a towering palm tree with a single trunk that can reach heights of up to 20 meters. The trunk is large, with a diameter of nearly a meter, and it widens significantly at the base, forming a bulbous and flaring shape. In Tunisia, *P. canariensis* is a naturalized alien species and is cultivated as a street tree [7]. Thus, it was demonstrated that *P. canariensis* fruits could be used to create new innovative and functional foods while also valuing a common byproduct [8].

In Tunisia, the presence of the RPW was first reported in the Carthage region on *Phoenix canariensis* [9]. About 30 to 40 Canary palm trees exhibited severe desiccation and significant apex damage. In 2014, Chihaoui [10] reported the death of approximately 254 palm trees [10]. Our latest data indicate that, by 2024, a total of 11,841 palm trees had been attacked across various regions including the capital Tunis, Bizerte, Manouba, parts of Nabeul, Zaghouan, and Bouficha. The most significant

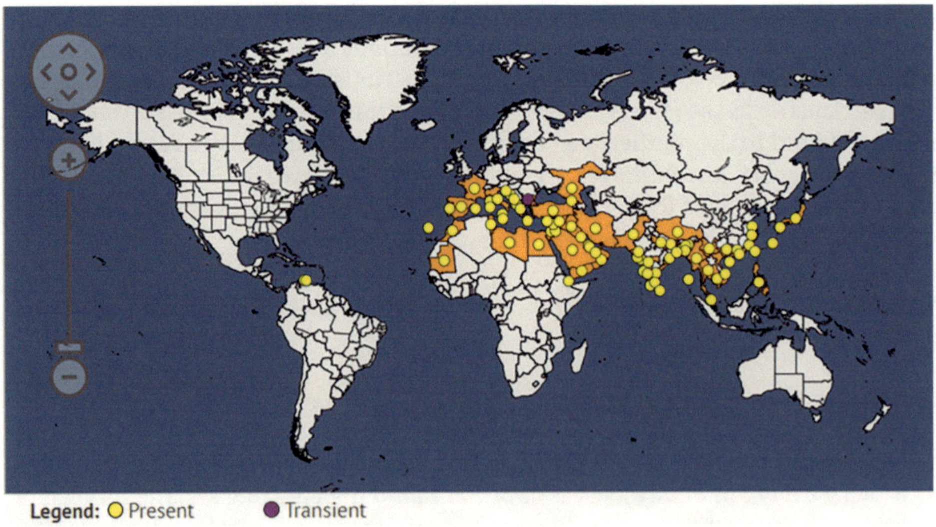

Figure 1.
EPPO global database 09–2023.

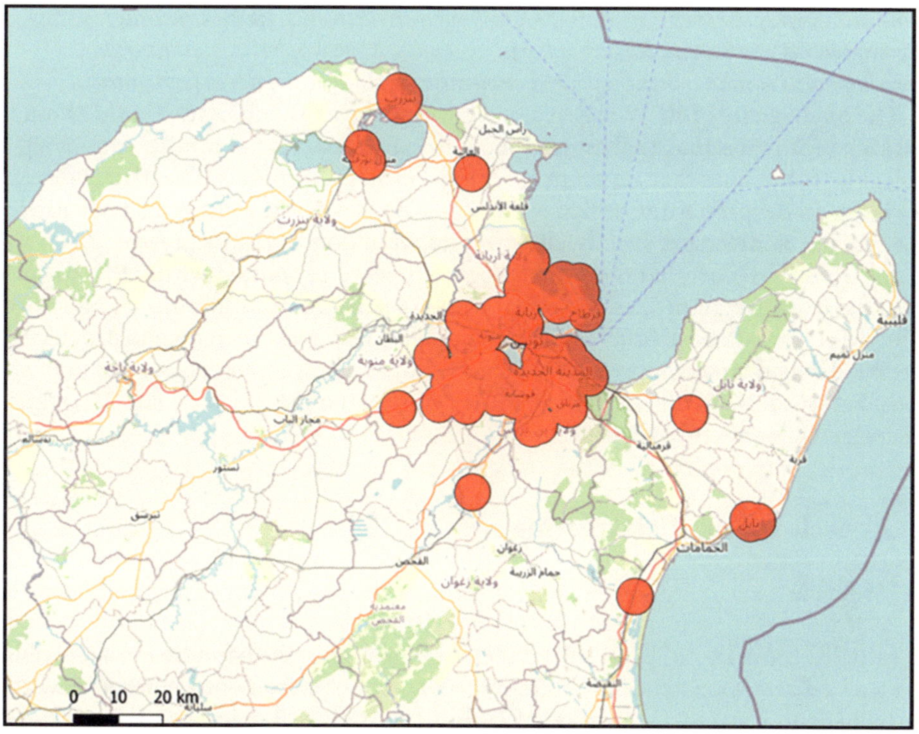

Figure 2.
Infested area in Tunisia (2024).

infestation peak occurred in 2016, with 2500 infested Canary palms. Since 2020, the average annual infestation rate has stabilized at around 829 trees per year (**Figure 2**).

Although the RPW is currently confined to the urban areas of Tunis, its presence represents a major socioeconomic threat to the date industry in the country's southern regions. On a global scale, the date palm serves as a primary source of export revenue; in 2022, Tunisia's exports accounted for 10.7% of the global date export values and generate significant income. It ranked fifth in the world [11]. The date palm is also among the crop species that sustains the livelihoods of millions of rural small-scale farmers. If the infestation of the RPW continues to spread to the south of the country, it will jeopardize local agriculture and the economy, both of which heavily depend on date production. Effective monitoring and management strategies are crucial to combat this invasive pest and protect the date palm industry in the region.

2. Taxonomy, biodiversity, and biology of the RPW

2.1 Taxonomy and biodiversity

The *Rhynchophorus* is the genus of the insect belonging to the family Dryophthoridae, the subfamily of Dryophthorinae, and the order Coleoptera that was initially described by Johann Friedrich Wilhelm Herbst in 1795. An evolutionary trait of Dryophthorine beetles is their exclusive feeding on monocot stems during both the larval and adult stages. To date, ten species of *Rhynchophorus* have been identified worldwide: *Rhynchophorus bilineatus, Rhynchophorus cruentatus, Rhynchophorus distinctus, Rhynchophorus labatus, Rhynchophorus palmarum, Rhynchophorus phoenicis, Rhynchophorus quadrangulus, Rhynchophorus richteri, Rhynchophorus vulneratus, and Rhynchophorus ferrugineus.*

The biodiversity of RPW was evaluated using two methodologies: firstly, through marking techniques that analyze morphometric patterns, and secondly, employing molecular techniques that augment traditional ecological studies. The dark color markings on the pronotum, which is the upper surface of the prothorax, were used to assess the biodiversity, specifically the number of dots or marks present. Sadder et al. [12] could report that there are 09 phenotypes of pronotum within *R. ferrugineus* collected from Kingdom Saoudi Arabia, while 24 patterns of pronotum markings were detected from southeast Asia [13]. Molecular tools however are known to provide precise, detailed, and comprehensive genetic information about species and populations. Few studies initially utilized the RAPD technique [14–17], COI [18–23], while others later adopted the ITS method [12].

3. Life cycle and artificial rearing

3.1 Life cycle

Initially, the male red palm weevil seeks refuge within the palm tree to avoid light and lead a discreet existence. By secreting aggregation pheromones synergistic with the palm odor, it attracts both sexes of *R. ferrugineus* [7, 8]. Typically, it is the females that initiate the red palm weevil's introduction into their plant host by drilling holes in the tender parts of the fronds, reaching the living tissues where they deposit their eggs [24]. About 2 to 3 days after mating, a female can lay between 200 and 400 eggs [25]. The young larvae hatch inside the living tissues, where they continue their development, causing considerable damage [26]. This larval development, representing the harmful phase, lasts from 1 to 4 months depending on seasonal conditions. Contrary to common assumption, prior injury is not necessary for red palm weevil infestation [27], although it may enhance infestations, as well as the presence of other pests such as different species of Rhinoceros beetles or rodents (**Figure 3**) [28].

The larvae burrow destructive galleries that damage the vascular system and induce rot. In the final larval stage, they construct cocoons for pupation [25]. A complete generation takes from 2 to 5 months [29], with typically 3 to 4 generations per year in France [4]. The red palm weevil overwinters at all developmental stages [29], and its minimum larval development temperature is 15°C. Lethal temperatures for egg, larva, and pupa stages are 10°C, 5°C, and – 2°C, respectively [30], while optimal conditions for egg hatching are 26°C and 64% relative humidity [31], and the insect remains active between 25 and 40°C [28]. The red palm weevil is capable of flying distances of 500 to 800 m over an unspecified period [13]. A study on the flight capability of *R. ferrugineus* adults, conducted during computer-guided flights, showed that 54% were classified as short distance (<100 m), 36% as medium distance (100–5000 m), and 10% as long distance (>5000 m) [32].

3.2 Artificial rearing

Artificial diets have demonstrated significant potential in providing a controlled and manipulable system for studying insect reproductive performance and establishing control methods. Initially, coconut was substituted with sugarcane, incorporated

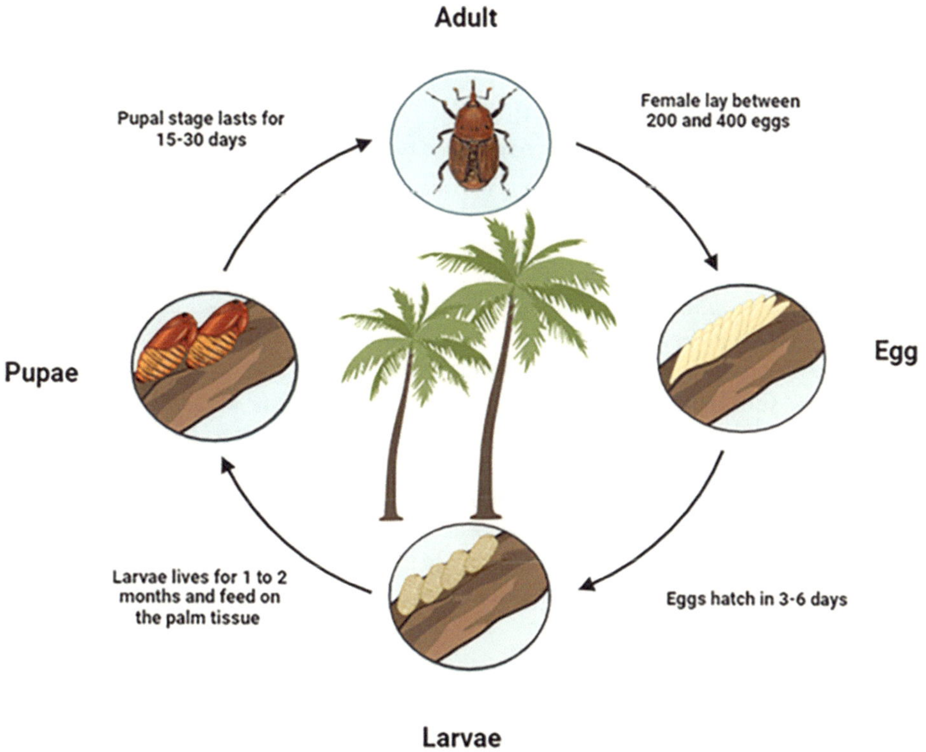

Figure 3.
Life cycle of the red Plam weevil.

into agar nutrient or in pieces, depending on the larval stage [33, 34]. Subsequent attempts incorporated mixtures containing coconut cake, sugarcane bagasse, wheat flour, corn flour, potatoes, yeast, sucrose, minerals, vitamins, preservatives, and a variety of other components [35–37].

Trials also utilized date palm derivatives for artificial diets. Date palm trunks were effectively used by Al-Ayedh [38], who later evaluated the development of RPW on four date palm cultivars—"Khalas," "Sukkary," "Khasab," and "Sillaj"—finding that "Sukkary" supported significantly better growth across all developmental stages [39]. This was corroborated by Aldawood and Rasool [40], who found that date palm cultivars significantly influenced RPW populations. Furthermore, RPW was successfully reared on a meridic diet, devoid of host components, composed of agar, distilled water, commercial yeast, brewer's yeast, wheat meal, corn flour, benzoic acid, ascorbic acid, sorbic acid, a vitamin mix, and tetracycline hydrochloride. This diet supported the complete life cycle of RPW without any deformities compared to those reared on natural hosts [41].

4. RPW geographical range and host plants

Until the 1970s, the red palm weevil was considered a tropical pest. However, from the 1980s and 1990s, its distribution range expanded toward the Middle East, where it was reported on date palms, notably in Egypt in 1992 [42]. This expansion was facilitated by commercial exchanges with Europe. Consequently, it was detected in Spain as early as 1993, in Andalusia and Valencia, then in Murcia and Catalonia between

2003 and 2006 [43]. Subsequently, RPW was spotted in Palestine in 1999 and in Italy in 2004 (in Tuscany in 2004, in Campania and Sicily in 2005). In the same year, the pest was introduced with infested palm trees into Turkey and Greece. In 2006, the insect invaded the French Riviera [44].

R. ferrugineus, affiliated with the Arecaceae family [45, 46], finds its primary host plant in Saudi Arabia, the United Arab Emirates, and Egypt, in *Phoenix dactylifera* L. (date palm) [47, 48]. Although generally susceptible, *P. dactylifera* and *P. canariensis* are preferred host plants [49]. However, other species such as *Areca catechu, Arenga pinnata, Borassus flabellifer, Caryota cumingii, Caryota maxima, Caryota nucifera, Corypha umbraculifera, Elaeis guineensis, Livistona chinensis, Livistonasaribus, Metroxylonsagu, Nypafruticans, Oncosperma horridum, Oncosperma tigillarium*, and *Roystonea regia* are also potential hosts for *R. ferrugineus* [46].

In Tunisia, four palm species are recorded, notably *Phoenix canariensis* L., Washingtonia filifera (Linden ex André) H. Wendl. 1879, *Chamaerops humilis*, and *Phoenix dactylifera* L. 1753, with the first two being predominant in the Greater Tunis region. However, the insect remains absent from the southern Tunisian oases. The Canary Island date palm, *Phoenix canariensis*, is widely preferred over the date palm, and the spread of the red palm weevilis closely linked to the trade of palm seedlings for ornamental purposes, particularly in large Mediterranean regions.

5. The ecological and economic impact of the red palm weevil in the Euro-Mediterranean region

The red palm weevil (RPW) has a profound impact on ecosystems, particularly in the Euro-Mediterranean region, where palms play a crucial role. As an invasive alien species (IAS), RPW's introduction through human activities has led to widespread disruption of local biodiversity and ecological balance [50–52]. The concept of ecosystem services, highlighted by the Millennium Ecosystem Assessment (MEA) in 2005, categorizes the benefits humans derive from ecosystems into provisioning, cultural, and regulating services. Palms, especially the Canary Island date palm (*Phoenix canariensis*), are integral to urban and rural landscapes in this region. They provide biotic materials, such as fiber and food, and serve as ornamental plants that significantly enhance the beauty of urban spaces. In cities, where green spaces are limited and often unevenly distributed, palms contribute to urban forests that improve the quality of life for residents and the experiences of tourists. They offer esthetic pleasure, increase property values, foster a sense of place and community, and reduce stress [53–57]. Beyond these cultural services, palms provide critical habitats for various wildlife species, thus supporting biodiversity. They play a role in regulating services by contributing to urban microclimates, aiding in temperature regulation, and improving air quality. The presence of palms and other green infrastructure can also promote health and fitness, spiritual well-being, and opportunities for outdoor learning and recreation [58]. The infestation of palms by RPW results in the loss of these vital ecosystem services. The esthetic and ecological damage caused by this pest necessitates significant public and private expenditure on control measures, including the felling and destruction of severely infested palms or their sanitization and restoration when possible. Recent studies, such as those by [59, 60], underscore the severe damage caused by RPW and highlight the need for integrated management strategies. These efforts emphasize the economic and ecological costs associated with RPW infestations and highlight the urgent need for effective management and preventive strategies to protect palm species and the valuable services they provide to both urban and rural environments [59, 60].

6. Chemical control of red palm weevil: their fate in the environment and impacts on biodiversity

Insecticides are commonly used to manage *R. ferrugineus* populations, applied through various methods such as sprays or paints on stipes, frond bases, and trunks, as soil and trunk injections, or as soil or crown drenches [61]. However, irrespective of their methods of use, insecticide residues can accumulate in various ecological compartments, including palm crops, soil, and aquatic ecosystems, which may result in environmental pollution, risks to human health and ecosystem integrity, as well as biodiversity conservation efforts [62–66].

Another significant limitation of using insecticides to control *R. ferrugineus* is their potential for nontarget effects on beneficial organisms and ecosystem services. Insecticides likely affect nontarget herbivores and beneficial biota that feed upon palm leaves, (extra-) floral nectar or pollen. Such organisms include insect or vertebrate pollinators and biological control agents. Pesticides can also harm soil microorganisms, such as mycorrhizal species and earthworms responsible for nitrogen fixation, organic matter decomposition, and water and nutrient absorption [67]. Pesticides, along with habitat loss and climate change, are major factors impacting biological diversity. Importantly, this impact is also manifested as a reduction of the potential of natural enemies to control pest organisms. They can cause toxic effects on directly exposed organisms in the short term, and over the long term, they may alter habitats and disrupt food chains [68]. Additionally, pesticide drift and runoff can contaminate nearby habitats and water sources, affecting nontarget organisms and ecosystem functions beyond palm plantations [69]. The intensive use of chemical pesticides to manage *R. ferrugineus* populations can contribute to environmental degradation, leading to biodiversity loss, soil health deterioration, and water quality decline [69]. The loss of natural habitats and biodiversity due to pesticide-intensive practices can reduce ecosystem resilience and the ability of ecosystems to provide essential services like pollination and pest regulation [70]. Biodiversity is crucial in regulating *R. ferrugineus* populations, as healthy ecosystems with diverse flora and fauna provide natural predators and competitors that help control *R. ferrugineus* [71].

Moreover, chemical pesticides pest control methods often exhibit limited long-term efficacy. Overuse of insecticides can result in resistance development, rendering them less effective or even ineffective in controlling outbreaks [68, 72]. Additionally, reliance on chemical pesticides can disrupt natural pest control mechanisms and exacerbate pest problems, leading to a cycle of dependence on increasingly intensive pest management practices. This dependence on chemical insecticides for both prevention and treatment highlights the need to explore alternative approaches. Research into the effectiveness of natural insecticides and management with natural enemies against *R. ferrugineus* is essential, advocating for their adoption and integration into control strategies [73].

7. Approaches of conservation of palms based on biological control

Conventional control methods, like using insecticides, have been mostly ineffective against the weevil because of its distinctive behavioral and biological

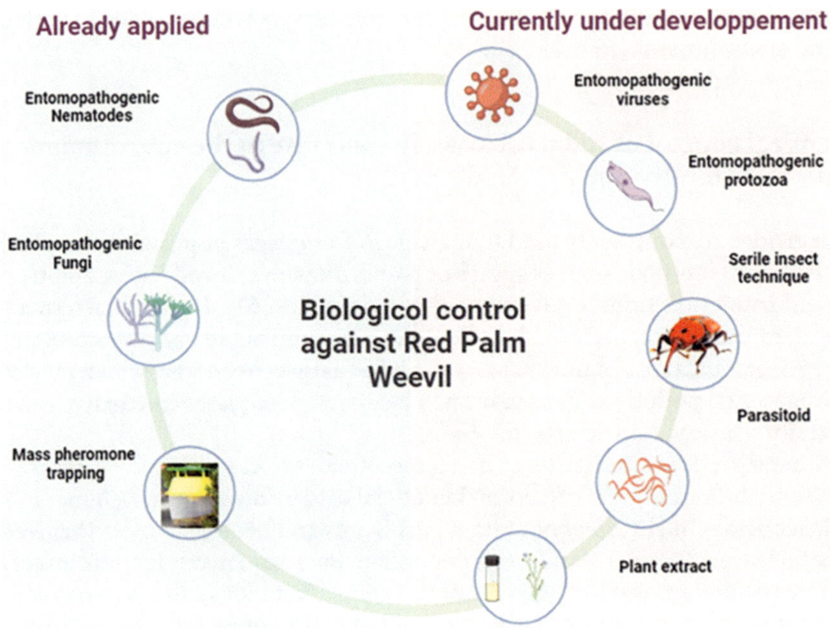

Figure 4.
Advancements in biological control of the red palm weevil: emerging research and established techniques.

traits. Therefore, biological control presents a viable alternative for managing the RPW. Various natural enemies, including entomopathogenic microorganisms such as fungi, nematodes, bacteria, viruses, and protozoa, have been reported to target *R. ferrugineus* (**Figure 4**).

7.1 Role of biological control of *R. ferrugineus* in conserving biodiversity

Since chemical applications elicit serious concerns related to environmental pollution and insects' resistance, and affect human health, and destroyed biodiversity, eco-friendly biological control has now attracted high interests. Biological control is a broad term including a variety of management actions based on the use of natural enemies of the invader. In the presence of natural enemies, pesticide use does not provide any advantage for long-term pest management [74]. Opportunities to harness this field-level biodiversity appear limitless: different strategies are available to conserve resident natural enemies, augment their field populations with laboratory-reared individuals, or judiciously translocate biological control agents from the pest's region of origin [75, 76]. By using biological control agent, the pest population can be effectively managed without the use of chemical pesticides. As a result, the palm crops are preserved, allowing the ecosystem to maintain its biodiversity. The presence of healthy palm crops supports a richer diversity of species, contributing to a more balanced and resilient ecosystem. In this way, biological control not only helps protect palm trees but also enhances the overall biodiversity of the ecosystem by preserving the natural habitat and the species that depend on it [71]. The biological control of *R. ferrugineus* could not be a utopian or a cost-consuming goal, especially when combined with other techniques aimed at achieving long-term, sustainable, and environmentally friendly control strategies.

7.2 Entomopathogenic fungi (EPF)

Entomopathogenic fungi (EPF) play a crucial role in the biological control of insects due to their parasitic nature, which enables them to infect and kill arthropods. As biopesticides, they offer a safer alternative to chemical insecticides. EPF operate through a distinct mechanism; they infect insects by penetrating their cuticle upon contact [77]. The infection can spread directly or indirectly through inoculum transmission, where spores are passed from infected insects or cadavers to untreated insects or future generations [78]. Their effectiveness in controlling the RPW has been tested in both laboratory and field conditions [79, 80]. Among the most studied EPF are *Beauveria bassiana* (Bals.-Criv.) Vuill. (1912) (Hypocreales: Cordycipitaceae) and *Metarhizium anisopliae* (Metschn.) Sorokīn. (1883) (Hypocreales: Clavicipitaceae). Studies have demonstrated that these fungi are potent biological control agents against *R. ferrugineus*, showing significant mortality rates across various developmental stages, including pupae, larvae, and adults [81]. These fungi have been observed to cause over 88% larval mortality and 75% adult mortality [82]. Additionally, they have caused 80–82% mortality in eggs and hatched larvae, with adult mortality reaching up to 100% within 2–3 weeks [83]. Field studies have demonstrated the effectiveness of *B. bassiana* in reducing RPW populations and mitigating damage in infested areas [84, 85].

7.3 Entomopathogenic bacteria (EPB)

For over a century, bacteria have been employed as biocontrol agents to manage pests, offering an environmentally friendly alternative to chemical pesticides for crop protection and improvement [86]. Entomopathogenic bacteria (EPB) include both spore-forming species such as Bacillus spp., Paenibacillus spp., and Clostridium spp., as well as non-spore-forming ones from the genera *Pseudomonas, Serratia, Yersinia, Photorhabdus, and Xenorhabdus* [87]. Among these, *Bacillus thuringiensis* is widely recognized as the most extensively employed bioinsecticide against several agricultural pests. *B. thuringiensis* has demonstrated effectiveness against numerous insect orders, including Coleoptera [87]. The effectiveness of *B. thuringiensis* stems from its insecticidal toxins, which are active during the infection process. Moreover, these bacteria generate a variety of virulence factors that enhance their ability to kill insects [88]. *B. thuringiensis* has shown promise as a biocontrol agent against *R. ferrugineus*, causing 94% mortality of the second larval instars after 15 days of treatment [89]. Another study demonstrated the efficacy of *B. thuringiensis* against *R. ferrugineus*, achieving 100% mortality after 72 hours of treatment. Other species from the Bacillaceae family also exhibit entomopathogenic activity against *R. ferrugineus,* such as *B. sphaericus, B. laterosporus,* and *B. megaterium*, with larval mortality rates ranging between 40% and 60% [90].

7.4 Entomopathogenic nematodes (EPNs)

Entomopathogenic nematodes (EPNs) are soil-dwelling organisms from the Steinernematidae and Heterorhabditidae families. They have garnered significant attention as biocontrol agents over the past few decades [91, 92]. EPNs enter insect hosts through natural openings such as the mouth, anus, or spiracles [93]. Once inside, EPNs release symbiotic bacteria into the host's intestine, which proliferate, causing septicemia and rapidly killing the host within 24 to 48 hours [94]. Research has demonstrated their efficacy in managing various insect pests, including the RPW [82]. Multiple studies suggest that EPNs could serve as a viable alternative to synthetic insecticides. For instance, *Steinernema carpocapsae* achieved a 98.9% mortality rate in RPW larvae 8 days posttreatment in laboratory trials, followed by *Heterorhabditis bacteriophora* at 86.9% and *Steinernema feltiae* at 38.9% [95]. Additionally, *Heterorhabditis bacteriophora* was found to cause a 94% mortality rate in larvae [96]. Under laboratory conditions, the Saudi strain *Heterorhabditi sindicus* caused 100%, 70%, and 75% mortality in young, grown larvae, and adults of the RPW, respectively. In field conditions, it achieved a mortality rate of 60% in larvae and 46% in adults in infested trees [97]. *Teratorhabditissy papillata* was also reported as a natural enemy of *R. ferrugineus*. Ibrahim et al. [98] found that *T. synpapillata* reduced the lifespans of *R. ferrugineus* larvae by 35–66%, while it had no effect on weevil adults. However, in Tunisia, the study by Loulou et al. [99] confirmed the results of Ibrahim et al. [98] and showed that *T. synpapillata* has no potential effect on RPW adults.

7.5 Parasitoids and predators

Numerous studies have identified various parasitoids and predators of RPW, including *Chelisoches morio* (Fabricius) (Dermaptera: Chelisochidae) [100], *Anisolabis maritima* Bonelli, *Euborellia annulipes* (Lucas), and *Xylocoris galactinum* (Fieber) [101, 102], as well as *Platymeris laevicollis* [101]. Other notable predators include *Sarcophaga fuscicauda* Bottcher [101, 103, 104], *Scolia erratica* Smith (Hym., Scoliidae) [8, 97], and *Paratheresia menezesi* Townsend [104, 105]. Additionally, various mites have been reported as parasites or predators of RPW [106–108]. The parasitoid mites *Rhynchopolypus rhynchophori*, *Rhynchopolipus swiftae*, and have been recorded parasitizing adult weevils [109, 110]. Slimane-Kharrat and Ouali [111] identified three mite species associated with RPW in Tunisia: *Centrouropoda almerodai*, attached to the underside of elytra; *Uroobovella marginata*, adhered to the pygidium, thorax, and head; and *Uroobovella javae*, fixed to the antenna, thorax, and legs.

7.6 Bioinsecticides (plant extracts)

Plant-derived biopesticides offer an eco-friendly and sustainable approach to controlling RPW, minimizing environmental pollution and human toxicity while safeguarding nontarget organisms [112]. These biopesticides include compounds such as azadirachtin from neem, renowned for its insecticidal properties in pest management [113]. Studies have demonstrated the effectiveness of extracts from *Calotropis procera* against early RPW larvae stages [114], while vegetable oils like Rio ipecac oil have shown varying levels of toxicity against RPW larvae and adults [115]. Additionally, extracts from *Ocimum basilicum* have been found effective against RPW adults and larvae, and cardamom and clove seed powders have shown high mortality rates in RPW under laboratory conditions [116]. These findings underscore the potential of plant-based biopesticides in integrated pest management strategies against *R. ferrugineus*.

7.7 Host plant resistance (HPR)

Host plant resistance (HPR) leverages the adaptive characteristics inherent in both wild and cultivated plant hosts, enabling them to defend against harmful herbivores such as the RPW. This resistance encompasses natural defense mechanisms, including making plants unpalatable, inhibiting insect growth due to nutrient scarcity, or providing immunity [117–120]. The population growth rate of insects, considering developmental time, survivorship, and fecundity, indicates the suitability of host species [121]. In Spain, studies have demonstrated that *Washingtonia filifera* and *Chamaerops humilis* possess antibiotic and antixenotic properties, respectively, against RPW, whereas *Phoenix canariensis* is highly preferred for the pest's development. Similar studies in China revealed that RPW development, survival, and reproduction were more favorable on *P. canariensis* and *W. filifera* compared to *P. sylvestris*, which was the least suitable among the five palm species evaluated in laboratory conditions. In Saudi Arabia, Al-Ayedh [38] found that laboratory-reared RPW had the longest lifespan on the date palm cultivar "Khalas," followed by "Sillaj," "Sukary," and "Khasab."

7.8 Viruses

Insect viral diseases have been recognized for centuries, but their use in pest control has gained prominence in recent decades [122]. Entomopathogenic viruses, belonging to 11 families, have been isolated from over a thousand species across at least 13 different orders of insects [123]. Research on viral infections in RPW is limited. Notably, a cytoplasmic polyhedrosis virus (CPV) has been reported to infect all developmental stages of RPW in India, causing malformed adults and significant suppression of the host population in late-larval stages [124]. El-Minshawy et al. [125] tested the efficacy of frozen-stored Rhynchophorus polyhedrosis virus at −4°C for 18 months and newly extracted Rhynchophorus polyhedrosis virus against larval instars of RPW, finding no significant differences in the LC50 of the two viruses. However, their potential as biocontrol agents is limited due to a lack of high virulence [96, 126].

7.9 Protozoa

Entomopathogenic protozoans, commonly known as microsporidians, are a highly diverse group of organisms that encompass thousands of species targeting invertebrates, including insects. They are usually host-specific and slow-acting, producing chronic infections with general debilitation of the host [127]. Protozoa, specifically flagellated protozoa, have been isolated from the hindgut of the RPW. These symbiotic protozoa play a crucial role in the weevil's digestive system, helping to break down complex nutrients and facilitating nutrient absorption [128]. The specific mechanisms by which these protozoa target the RPW are not fully understood, but their symbiotic relationship with the weevil highlights the complex interactions between insects and their gut microbiota [128]. Further research on these protozoa could provide valuable insights into the biology and ecology of the RPW, as well as potential strategies for managing this invasive pest.

7.10 Can biological control of *R. ferrugineus* cause loss of biodiversity?

Biological control, while generally considered more environmentally friendly than chemical control, can potentially lead to a loss of biodiversity if not carefully managed. This risk primarily arises when non-native is introduced as biological control agents to target specific pests. If these introduced species become invasive, they can disrupt local ecosystems, outcompete or prey on native species, and reduce biodiversity [129]. For example, a biological control agent might target not only the intended pest but also non-target organisms that play important roles in the ecosystem. This can lead to unintended consequences, such as the decline or extinction of native species, disruption of food webs, and altered ecosystem functions. However, host-range tests and risk assessments of the most promising agents should be given high priority to avoid any potential-post-release threats to non-target species in introduced areas [75, 101]. To date, no research was interested to the effects of the use of biological control agents against *R. ferrugineus* on biodiversity. While the biological control of *R. ferrugineus* carries risks for biodiversity, these can be mitigated through careful planning, targeted application, and continuous monitoring to minimize non-target impacts. When properly researched and carefully implemented, biological control can be an effective and sustainable method for pest management with minimal impact on biodiversity. The key is to thoroughly assess the potential risks and ensure that the control agents are specific to the target pests and will not harm non-target species or disrupt the local ecosystem [101]. However, future success will largely depend on increased investment in research and development by governments and related organizations committed to reducing reliance on chemical control methods [75].

8. Omics research and technologies: progress and application to mitigate of palm weevil *R. ferrugineus* damage and maintaining ecosystem health and biodiversity

Recent advances in omics technologies, including transcriptomics, genomics, metabolomics, proteomics, and microbiomics, have significantly enhanced researchers' understanding of insect pest biology and behavior, particularly in host–pest interactions, thus enabling the development of more effective management strategies [130]. Studying palm weevil interactions with their host palms is of particular interest, given the significant threat this pest poses to agricultural biodiversity and food security [131]. By employing omics approaches, researchers can gain insights into the genetic and biochemical processes underlying the palm weevil–host palm relationship. This interdisciplinary approach has the potential to revolutionize our understanding of these interactions and pave the way for innovative solutions to mitigate the impact of palm weevil infestations on palm plantations [132–134]. Recent years have seen significant advancements in omics research on *R. ferrugineus* [135]. Since 2013, transcriptome studies have been conducted to identify genes specifically expressed during the weevil's development and infestation of palm trees [136, 137]. A notable achievement in this field was the study by Wang et al., [136], which performed the first large-scale *de novo* cDNA library sequencing for *R. ferrugineus*, annotating the results using known insect genes and the *Tribolium castaneum* genome assembly. The latest genome assembly (chromosomal-level assembly) is around 779 Mb in size, with an N50 of ~43 Mb, consistent with previously reported genome size estimates [138, 139]. The proposed genome assembly is significantly superior to existing assemblies in terms of contiguity, integrity, and genome completeness [131, 138, 140]. This high-quality genome assembly is a valuable resource that will contribute to understanding *R. ferrugineus* diversity, genomic variation, molecular evolution, and environmental adaptation, which could eventually lead to molecular genetic-based pest control [133–135].

Considering olfaction's key role in *R. ferrugineus* chemical ecology, several authors performed transcriptomics studies on *R. ferrugineus* and identified putative genes and proteins related to the chemosensory system [141, 142]. This information is a valuable resource that enables analyses to identify potential targets for preventing the spread of *R. ferrugineus* and enhanced pest management strategies [141]. For example, using RNAi analysis, several studies were performed to inhibit of genes or proteins related to the chemosensory system of *R. ferrugineus* [141, 143].

However, to effectively develop targeted pest control strategies to control *R. ferrugineus*, it is crucial to emphasize the need for a deeper understanding of its interaction with its plant host. In this context, analyzing the volatilomics and metabolomics of palm trees infested by *R. ferrugineus* can reveal potential biomarkers for early detection and monitoring of weevil populations and allows identification gene expression affecting antioxidative defense responses [144, 145]. Proteomics research provides insight into production of defensive secondary metabolites by palms that may have use in developing next-generation biopesticides [146]. Signature compounds identified from volatilomics, metabolomics, and proteomics research may help identify opportunities for control of this pest [147].

Other omics technologies may also contribute to the development of strategies to control *R. ferrugineus*. These researches were interested to characterization of bacterial communities associated with *R. ferrugineus* and its hosts. Investigating the weevil's gut microbiome provides insights into symbiotic relationships that can be disrupted to affect the weevil's health and survival. Improved understanding of the identities and functions of microbial symbionts may provide insights into coevolutionary relationships with palm weevils and processes affecting weevil interactions with palm hosts. For example, Alshammari et al. [148], through 16S rRNA gene sequencing by Illumina MiSeq, compared the microbial communities of infected and uninfected palm trees in Northern Saudi Arabia. Myxococcota, Acidobacteriota, and Firmicutes were the dominant phyla in *R. ferrugineus*-infected tree trunk, and Pseudomonadaceae was the most prominent family. This research also presents an opportunity to identify biological control agents and/or plant growth-promoting bacteria or fungi that can be harnessed or engineered to target the weevil and offer environmentally friendly pest control solutions.

Moving forward, further studies integrating omics approaches with traditional breeding and management strategies will be essential to develop sustainable and effective pest control measures that can mitigate the economic and ecological impacts of palm weevil infestations. Omics technologies offer powerful tools for understanding and mitigating the impact of the red palm weevil on palm trees. By integrating genomics, transcriptomics, proteomics, metabolomics, and microbiomics, we can develop targeted, sustainable, and effective pest management strategies. These approaches help in not only controlling weevil populations but also maintaining ecosystem health and biodiversity, ensuring the long-term preservation of palm tree ecosystems.

9. Potential application of the sterile insect technique (SIT) against the RPW

The sterile insect technique (SIT) is an environment-friendly method of pest control that integrates well into area-wide integrated pest management (AW-IPM) programs. The SIT consists in the mass production, sterilization, and release of insects into infested areas, where sterile males mate with wild females resulting in no offspring and thus reducing pest populations [149]. The SIT could improve the likelihood of successful RPW control, especially considering physically and ecologically isolated or susceptible areas [150, 151]. This technique can be applied alone or combined with other control methods. Several studies [152–154] have demonstrated the feasibility of the SIT to control RPW, but certain reproductive aspects of this species, such as polyandry and post-copulative sperm competitiveness in females, could reduce the feasibility of this technique for its control [151, 155, 156]. Several studies have demonstrated that radiation doses ranging from 15 to 80 Gy significantly reduce sperm quality and subsequently affect egg hatching [153, 157, 158]. For instance, in India, an estimated ratio of 10 sterile males to 1 wild male was required to decrease population densities [153].

10. Approaches to conserving biodiversity of the canary palm trees in Tunisia

After the accidental introduction of the red palm weevil (RPW) and its initial detection, the strategy began with the application of a foliar treatment using chlorpyrifos-ethyl three times upon detecting symptoms on the palm tree (stage 1 or 2), which included misalignments and notches. RPW infestation in date palms was categorized on a scale of 0 to 5, with zero representing uninfected palms and five indicating advanced stages of infestation [159]. Standard traps, such as ground rat traps with kairomone and pheromone, were employed to indicate the presence of infested palm trees within a radius of approximately 300 meters, necessitating the identification of the infested tree. In case of detection of a new infested palm, it was subjected to immediate sanitation that consists on chemical treatments to kill all the RPW present in the infested palm before and after the mechanical intervention.

Following a peak of infestation in 2016, a systematic foliar treatment approach was implemented for all healthy and infested palm trees within the visual detection area (**Figure 5**). A mass trapping trial was conducted using 2000 traps (5 traps/ha), which required intensive management and was not pursued. Endotherapy was also tested, with emamectin benzoate injected once a year at 1 meter above the soil to reach the vascular system of the Canary palm tree (50 to 60 cc per palm tree). It has the advantage of significantly minimizing health and environmental risks since the insecticide is confined within the palm tree. It is practical and inexpensive to implement. Imidacloprid was similarly used in both endotherapy and foliar treatments.

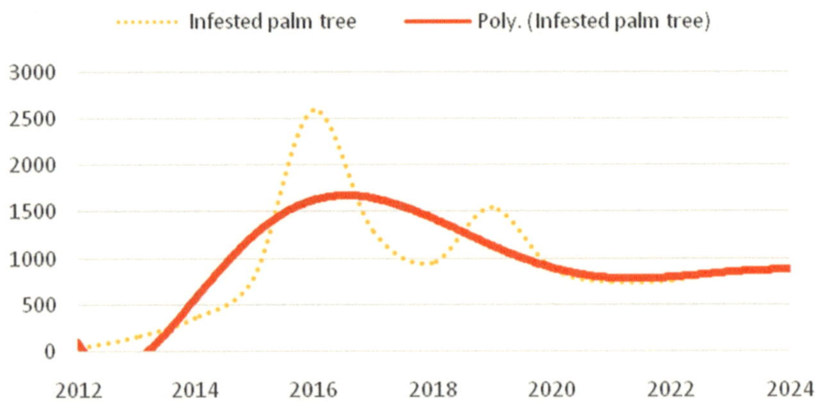

Figure 5.
Evolution number of infested P. canariensis *palm trees.*

The treatment frequency consisted of one injection per year in the spring, with an additional injection sometimes administered in the autumn. Unfortunately, there was no Canary tree inventory that could help rapid and timely inspection of the first symptoms.

11. Regulatory issues and attempts to face Tunisian oasis infestation

Law No. 92–72 of August 3, 1992, concerning the revision of the legislation related to plant protection, establishes by decree the list of quarantine pests against which control is mandatory (https://faolex.fao.org/docs/pdf/tun11737.pdf). Additionally, it specifies by decree the list of plants and plant products whose entry into Tunisian territory is prohibited. This list was updated by a decree from the Ministry of Agriculture dated June 26, 2015, establishing the list of plants and plant products whose entry into Tunisian territory is prohibited (http://www.agridata.tn/dataset/arrete-du-ministre-de-lagriculture-du-26-juin-2015/resource/3de01d89-4c27-47ab-a517-19b3bdd199c5). This new list includes plants or plant products and articles made entirely or partially from palm species, except for coconut palm (*Cocos nucifera*) used for food after sterilization. It is also important to note that this law mentions compensatory indemnities that can be granted to the operator following the destruction of plants or plant products upon the decision of the phytosanitary inspector. This law can therefore be considered in establishing the date palm protection strategy in southern Tunisia. Since Tunisia is the largest supplier of dates to the European Union [160], a series of measures must be undertaken to address a potential intrusion of the red palm weevil into Tunisian oases. The Ministry of Agriculture has established observatories in the south that will carry out monitoring through trapping (1 trap/ha) and visual observation during spring and autumn. Immediate intervention is required; indeed, the strategy adopted by Morocco and Mauritania with strict quarantine laws has restricted the weevil to the original infestation sites and prevented its spread. The weevil was accidentally introduced into Mauritania through offshoots from the United Arab Emirates, with only one oasis declared infested. A rapid implementation of the eradication strategy was carried out with the mobilization of all stakeholders, with oasis owners playing a key role, leading to the successful eradication of the weevil [161].

As can be seen, the key to successfully eradicating the red palm weevil lies in quick and early action. It is essential to provide IoT tools that are programmed to perform tasks while minimizing human effort. The IoT tools will allow the real-time and remote control of the plant protection devices and the intelligent management of the data within the oasis [162].

12. Impact of *R. ferrugineus* infestations on palm agrosystem wildlife diversity

Palm tree agroecosystems have become vital habitats for local fauna, offering numerous ecosystem functions such as providing food and shelter for various animal species. These systems contribute to the conservation of biological and genetic diversity and support evolutionary processes [163]. Research on the ecological functions of palm species has primarily focused on their natural native ranges or has been heavily biased toward managed date palm plantations and fauna species considered pests, including non-arthropod pests [164]. However, data on the impact of *R. ferrugineus* infestations on palm agroecosystem wildlife diversity are lacking.

Numerous studies indicate that when alien insect species invade an ecosystem, they can disrupt the natural balance, reduce biodiversity, degrade habitats, alter 0native genetic diversity, and potentially transmit exotic diseases to native species [165, 166]. The infestation of *R. ferrugineus* likely creates significant ecological disruption by directly destroying palm species and indirectly affecting the broader ecosystem and its biodiversity [167, 168]. *R. ferrugineus* larvae bore into the trunks of palm trees, causing structural damage that often results in the tree's death. This not only leads to substantial economic losses, especially in regions dependent on palm products like dates, coconuts, and ornamental palms, but also impacts the broader ecosystem. The destruction of palm trees disrupts habitats for various species of birds, insects, and other wildlife, leading to a decline in these populations [163, 169]. Additionally, the removal of palm trees can alter local landscapes and affect ecosystem dynamics, potentially increasing erosion and changing soil composition. The loss of palms can also disturb the ecological balance, impacting the growth and survival of other plant species that coexist with them. Consequently, *R. ferrugineus* infestations threaten not just palm species but also the broader ecosystem, causing a significant reduction in biodiversity and ecological imbalance. During the initial stages of pest infestation, implementing early preventative and curative treatments is crucial. These measures help protect the palm from further damage and aid in its recovery from the infestation. Early intervention can significantly reduce the impact of pests, ensuring the health and longevity of palm trees while maintaining the balance of the ecosystem [170]. While synthetic pesticides have traditionally been relied upon for RPW control [171], their use is unsustainable and poses risks to the environment, biodiversity, and human health [64, 65].

13. Management tactics of *R. ferrugineus* and possible roles of wildlife diversity in palm agrosystem management

The effective management of *R. ferrugineus* requires an integrated strategy encompassing prevention, early detection, and rapid response [172, 173]. Integrated Pest Management (IPM) involves regular monitoring and early detection through routine inspections, pheromone traps, and other tools. Implementing proactive surveillance programs that utilize new technologies in vulnerable areas allows for rapid detection and containment, increasing the likelihood of eradicating early-stage populations and enhancing control efforts in commercial production areas [61].

Biological control methods, such as introducing natural predators or parasites, can help manage weevil populations without harming other wildlife. Mazza et al. [101] identified over 50 natural predators, including vertebrates, insects, mites, yeast, nematodes, fungi, bacteria, and viruses, that target *R. ferrugineus*. Judicious application of insecticides targets affected areas, ensuring minimal environmental impact [174, 175]. Insecticides can be applied prophylactically to protect uninfested palms or curatively to kill weevils during early infestation stages. These procedures should be carried out with caution and by trained professionals or under technical supervision to ensure effectiveness and minimize potential harm to the environment and nontarget species [170].

Cultural practices, such as proper sanitation, removing infested plant material, and planting resistant palm varieties, also reduce the likelihood of infestation [176, 177]. Habitat conservation efforts, including habitat restoration and the creation of biodiversity corridors, support wildlife populations by replanting native vegetation and maintaining pathways for wildlife movement. Implementing restoration programs to replant and rehabilitate affected areas can help recover palm tree populations. These programs should consider genetic diversity to enhance resilience against future threats.

Public awareness and education initiatives, such as community engagement and training programs for farmers, landscapers, and gardeners, emphasize the importance of early detection and proper management practices [178, 179]. Farmer participation and cooperation are vital for effective pest management in area-wide operations [180]. Increasing farmers' involvement in managing *R. ferrugineus* on their farms and ensuring government support for the program can enhance engagement and participation.

Policy and regulation measures, including quarantine enforcement and funding for research, are critical for preventing the spread of *R. ferrugineus* and supporting conservation efforts. Ongoing research and the use of innovative technologies, such as drones and remote sensing, enhance early detection and monitoring of weevil populations [181]. By integrating these strategies, we can effectively manage *R. ferrugineus* infestations, protect wildlife, preserve natural habitats, and maintain biodiversity in affected areas.

14. Conclusion

An integrated strategy combining prevention, biological control, cultural practices, habitat conservation, public engagement, and technological innovation is essential for managing *R. ferrugineus* infestations and conserving biodiversity. By adopting these approaches, we can protect palm tree populations, support ecological balance, and ensure the sustainability of affected ecosystems.

Author details

Meriem Msaad Guerfali[1*], Kamel Charaabi[1], Mohamed Chaouachi[2], Sana Bouzembila[1], Salma Fadhl[1], Mariam Boudabouz[1] and Haythem Hamden[1]

1 Laboratory of Biotechnology and Nuclear Technologies, LR16CNSTN01, National Centre of Nuclear Sciences and Technologies, Sidi Thabet, Tunisia

2 Direction Générale de la Santé Végétale et du Contrôle des Intrants Agricoles (DGSVCIA), Tunis, Tunisia

*Address all correspondence to: msaad_tn@yahoo.fr

References

[1] Abraham VA, Abdulla-Koya M, Kurian C. Evaluation of seven insecticides for control of red palm weevil *Rhynchophorus ferrugineus* Fabr. Journal of Plantation Crops. 1975;**3**:71-72

[2] European and Mediterranean Plant Protection Organization (EPPO). Data sheets on quarantine pests. *Rhynchophorus ferrugineus*. EPPO Bulletin. 2008;**38**:55-59

[3] AlDosary N, AlDobai S, Faleiro J. Review on the management of red palm weevil *Rhynchophorus ferrugineus* Olivier in date palm *Phoenix dactylifera* L. Emirates Journal of Food and Agriculture. 2016;**28**:34

[4] Dembilio Ó, Jacas JA. Basic bio-ecological parameters of the invasive red palm weevil, *Rhynchophorus ferrugineus* (Coleoptera: Curculionidae), in *Phoenix canariensis* under Mediterranean climate. Bulletin of Entomological Research. 2011;**101**:153-163

[5] European and Mediterranean Plant Protection Organization (EPPO). 2023. Available from: https://gd.eppo.int/taxon/RHYCFE/distribution

[6] Spennemann Dirk HR. Epinastic growth and accordion-type leaf formation of a *Phoenix Canariensis* following herbicide application. Egyptian International Journal of Palms. 2023;**3**(1):69-83

[7] Sosa PA, Bouza N, Cabrera-Pérez MA, Luque A. Genetic variation in *Phoenix canariensis* and *P. Dactylifera* (Arecaceae) populations of gran Canaria using isoezyme electroforesis. Boletim do Museu Municipal do Funchal. 1998;**5**:443

[8] Turki M, Barbosa-Pereira L, Bertolino M, Essaidi I, Ghirardello D, Torri L, et al. Physico-chemical characterization of Tunisian canary palm (*Phoenix canariensis* Hort ex Chabaud) dates and evaluation of their addition in biscuits. Foods. 2020;**9**:695. DOI: 10.3390/foods9060695

[9] Chebbi H. First record of *Rhynchophorus ferrugineus* on *Phoenix canariensis* in Tunisia, Tunis. Journal of Plant Protection. 2011;**6**:149-154

[10] Chihaoui S. Evaluation de la technique de lutte par endothérapie contre le charançon rouge Rhynchophorus ferrugineus (Olivier) 1790 sur les palmiers d'ornement Phoenix canariensis (Chabaut) 1882 dans la zone de Grand Tunis. Tunisia: Mastère Bioagresseurs et Santé Végétale. Institut National Agronomique de Tunisie, Université de Carthage; 2014. 99 p

[11] ONAGRI. Observatoire National de l'Agriculture. 2023. Available from: http://www.onagri.tn/articles?id=5

[12] Sadder MT, Vidyasagar PS, Aldosari SA, Abdel-Azim MM, Al-Doss AA. Phylogeny of red palm weevil (*Rhynchophorus ferrugineus*) based on ITS1 and ITS2. Oriental Insects. 2015;**49**:1-14

[13] Wattanapongsiri A. A revision to the genera Rhynchophorus and Dynamis (Coleoptera: Curculionidae). Oregon State University. 1965;**1**:1-328

[14] Abulyazid I, Kamel IKE, Sharawi FA, El-Bermawi S. Comparison between different populations of red palm weevils *Rhynchophorus* species using RAPD-PCR. Journal Egyptian-German Society of Zoology. 2002;**38**:1-15

[15] Salama HS, Saker MM. DNA fingerprints of three different forms of red palm weevil collected from Egyption date palm orchards. Archives of Phytopathology and Plant Protection. 2002;**35**:299-306

[16] Gadelhak GG, Enan MR. Genetic diversity among populations of red palm weevil, *Rhynchophorus ferrugineus* Olivier (Coleoptera: Curculionidae), determined by random amplified polymorphic DNA-polymerase chain reaction (RAPD-PCR). International Journal of Agricultural Biology. 2005;**7**:395-399

[17] Al-Ayied HY, Alswailem AM, Shair O, Al Jabr AM. Evaluation of phylogenetic between three phenotypically different forms of red date palm weevil *Rhynchophorus ferrugineus* Oliv. Using PCR-based RAPD technique. Archives of Phytopathology and Plant Protection. 2006;**39**:303-309

[18] Song H, Buhay JE, Whiting MF, Crandall KA. Many species in one: DNA barcoding overestimates the number of species when nuclear mitochondrial pseudogenes are coamplified. National Academy of Sciences of the United States of America. 2008;**105**(36):13486-13491

[19] Derycke S, Vanaverbeke J, Rigaux A, Backeljau T, Moens T. Exploring the use of cytochrome oxidase c subunit 1 (COI) for DNA barcoding of free-living marine nematodes. PLoS One. 2010;**5**(10):e13716

[20] El-Mergawy RAAM, Al Ajlan AM, Abdalla N, Vassiliou V, Capdevielle-Dulac C, Kontodimas DC, et al. Preliminary study on geographical variation of cytochrome b gene and ITS2-rDNA among populations of *Rhynchophorus ferrugineus*. Journal of Agricultural Science and Technology B. 2011;**1**:189-197

[21] El-Mergawy RAAM, Faure N, Nasr MI, Avand-Faghih A, Rochat D, Silvain J-F. Mitochondrial genetic variation and invasion history of red palm weevil, *Rhynchophorus ferrugineus* (Coleoptera: Curculionidae), in middle-east and Mediterranean basin. International Journal of Agriculture and Biology. 2011;**13**:631-637

[22] Rugman-Jones PF, Kharrat S, Hoddle MS, Stouthamer R. The invasion of Tunisia by *Rhynchophorus ferrugineus* (Coleoptera: Curculionidae): Crossing an ocean or crossing a sea? The Florida Entomologist. 2017;**100**(2):262-265. Available from: http://www.jstor.org/stable/26358882

[23] Fouda MMA, Tufail M, Takeda M, et al. DNA barcoding and population genetic structure of the red palm weevil, *Rhynchophorus ferrugineus* (Coleoptera: Curculionidae) in Egypt based on mtDNA sequencing. Biologia. 2022;**77**:1017-1025. DOI: 10.1007/s11756-022-01033-7

[24] Ferry M, Gómez S. La stratégie de la dernière chance pour sauver les palmiers. Princeps. 2015;**1**:125-134

[25] Dembilio O, Llacer E, Martinez M, Jacas JA. Field efficacy of imidacloprid and Steinernema carpocapsae in a chitosan formulation against the red palm weevil *Rhynchophorus ferrugineus* (Coleoptera: Curculionidae) in *Phoenix canariensis*. Pest Management Science: Formerly Pesticide Science. 2011;**66**:365-370

[26] Giblin-Davis RM, Faleiro JR, Jacas JA, Peña JE, Vidyasagar PSPV. Biology and management of the red palm weevil, *Rhynchophorus ferrugineus*. In: Peña JE, editor. Potential Invasive Pests of Agricultural Crops. CABI Invasive Speciesseries N°3; 2013. pp. 1-34 [Published online]

[27] Ferry M, Cousin R, Chabernaud D, Ferrero F. An effective strategy to obtain very rapidly the red palm weevil decline in an area planted with ornamental palms. Arab Journal of Plant Protection. 2019;**37**(2):188-197

[28] Faleiro JR. Test and Refinery Protocols for Weevil Red Palm (RPW) Region-Wide Management, *Rhynchophorus Ferrugineus* (Olivier) as Agroecosystems in Al-Hassa, Saudi Arabia. Final Report. Al Hassa, Saudi Arabia: Palm Date Research Center. King Faisal University; 2009. 36 p

[29] Rochat D, Chapin E, Ferry M, Avand-Faghih A, Brun L. Le charançon rouge du palmier dans le bassin méditerranéen. Phytoma-La Défense des Végetaux. 2006;**595**:20-24

[30] Molina M, García CT. Biología y ecología del curculiónido rojo de la palmera [thesis]. 2005

[31] Salama HS, Abd-Elgawad MM. Isolation of heterorhabditid nematodes from palm tree planted areas and their implications in the red palm weevil control. Journal of Pest Science. 2001;**74**(2):43-45

[32] Ávalos JA, Martí-Campoy A, Soto A. Study of the flying ability of *Rhynchophorus ferrugineus* (Coleoptera: Dryophthoridae) adults using a computer-monitored flight mill. Bulletin of Entomological Research. 2014;**104**:462-470

[33] Rahalkar GW, Harwalkar MR, Rananvare HD. Development of red palm weevil, Rhyn- chophorus ferrugineus Oliv. On sugarcane. Indian Journal of Entomology. 1972;**34**:213-215

[34] Rananavare HD, Shantarum K, Harwalkar MR, Rahalkar GW. Methods for laboratory rearing of red date palm weevil, *Rhynchophorus ferrugineus* Oliv. Journal of Plantation Crops. 1975;**3**:65-67

[35] Rahalkar GW, Tamhankar AJ, Shantaram K. An artificial diet for rearing red palm weevil *Rhynchophorus ferrugineus* Oliv. Journal of Plantation Crops. 1978;**6**:61-64

[36] Kaakeh WAA, Khamis A, Aboul-Nour MM. The Red Palm Weevil: The Most Dangerous Agricultural Pest. Sharjah, United Arab Emirates: UAE University Printing Press; 2001. pp. 90-165

[37] El-Sebay Y, El-Lattef MAK, Makhlouf TM. Laboratory rearing of red palm weevil *Rhynchophorus ferrugineus* Oliv. (Coleoptera: Cur- culionidae) on artificial diet. Egyptian Journal of Agricultural Research. 2003;**81**:551-554

[38] Al-Ayedh H. Evaluation of date palm cultivars for rearing the red date palm weevil, *Rhynchophorus ferrugineus* (Coleoptera: Curculionidae). Florida Entomology. 2008;**91**:353-358

[39] Al-Ayedh HY, Rasool KG. Sex ratio and the role of mild relative humidity in mating behaviour of red date palm weevil *Rhynchophorus ferrugineus* Oliv. (Coleoptera: Curculionidae) gamma-irradiated adults. Journal of Applied Entomolology. 2009;**134**:157-162

[40] Aldawood AS, Rasool KG. Rearing optimization of red palm weevil: Rhynchophorus ferrugineus (Coleoptera: Curculionidae) on date palm: Phoenix dactylifera. Florida Entomologist. 1 Dec 2011;**94**(4):756-760

[41] El-Shafie HAF, Faleiro JR, Abo-El-Saad MM, Aleid SM. A meridic diet for laboratory rearing of red palm weevil, *Rhynchophorus ferrugineus* (Coleoptera: Curculionidae). Scientific Research and Essays. 2013;**8**(39):1924-1932

[42] Saleh MRA. Red palm Weevil, *Rhynchophorus ferrugineus* (Oliv.). The first record for Egypt and indeed the African Continent, List No. 10634 Africa, Collection No. 22563. British Museum Report of International Institute of Entomology. 1992; 56 Queen's Gate, London, SW 75 JR UK. p. 1

[43] Jacas JA, Dembilio Ó, Llacer E. Research activities focused on management of red palm weevil at the UJI-IVIA associated unit (region of Valencia, Spain). EPPO Bulletin. 2011;**41**(2):122-127

[44] EPPO. *Rhynchophorus ferrugineus* and *Rhynchophorus palmarum*. OEPP/EPPO Bulletin. 2007;**37**:571-579

[45] Thompson RT. Observations on the morphology and classification of weevils (Coleoptera, Curculionoidea) with a key to major groups. Journal of Natural History. 1992;**26**(4):835-891

[46] Rochat D, Dembilio O, Jaques AJ, Suma P, La Pergola A, Hamidi R, et al. *Rhynchophorus ferrugineus*: Taxonomy, Distribution, Biology, and Life Cycle Handbook of Major Palm Pests. 2017. pp. 69-104

[47] Abdel-Gawad MM. Date palm red weevil and pest control methods. Agricultre and Development in the Arab World-AOAD (Sudan). 1996;**15**:36-45

[48] Mahmud AI, Farminhao J, Viez ER. Red palm weevil (*Rhynchophorus ferrugineus* Olivier, 1790): Threat of palms. Journal of Biological Sciences. 2015;**15**(2):56-67

[49] Yang T-H, Li-Hsin W, Liao C-T, Li D, Shin TY, Kim JS, et al. Entomopathogenic fungi-mediated biological control of the red palm weevil *Rhynchophorus ferrugineus*. Journal of Asia-Pacific Entomology. 2023;**26**(1):1-6. DOI: 10.1016/j.aspen.2023.102037

[50] Vitousek PM, D'Antonio CM, Loope LL, Rejmánek M, Westbrooks R. Introduced species: A significant component of human-caused global change. New Zealand Journal of Ecology. 1997;**21**(1):1-16

[51] Hulme PE. Trade, transport and trouble: Managing invasive species pathways in an era of globalization. Journal of Applied Ecology. 2009;**46**(1):10-18. DOI: 10.1111/j.1365-2664.2008.01600.x

[52] Roy HE, Adriaens T, Isaac NJ, Kenis M, Onkelinx T, San Martin G, et al. Invasive alien predator causes rapid declines of native European ladybirds. Diversity and Distributions. 2010;**18**(7):717-725. DOI: 10.1111/j.1472-4642.2010.00823.x

[53] Deng J, Li J, Shen S. Evaluation of urban green spaces in China: A case study of Beijing. Chinese Geographical Science. 2010;**20**(2):145-156. DOI: 10.1007/s11769-010-0145-1

[54] Majumdar S, Altman I, Flynn CP. The Social and Cultural Contexts of Environmental Education. New York: Routledge; 2011. ISBN: 9780415808787

[55] Chaparro L, Terradas J. Ecological Services of Urban Forest in Barcelona. Centre de Recerca Ecologica i AAplicacions Forestals. Bellatera, Span: Universitat Autonoma de Barcelona; 2009. Available from: http://www.creaf.uab.es/

[56] Nowak DJ. Urban biodiversity and climate change. In: Urban Biodiversity and Design. Wiley-Blackwell; 2010. pp. 101-118. ISBN: 978-1-4051-7613-5

[57] Escobedo FJ, Kroeger T, Wagner JE. Urban forests and pollution mitigation:

Analyzing ecosystem services and disservices. Environmental Pollution. 2011;**159**(8-9):2078-2087. DOI: 10.1016/j.envpol.2011.01.010

[58] Tzoulas K, Korpela K, Venn S, Yli-Pelkonen V, Kazmierczak A, Niemelä J, et al. Promoting ecosystem and human health in urban areas using green infrastructure: A literature review. Landscape and Urban Planning. 2007;**81**(3):167-178. DOI: 10.1016/j.landurbplan.2007.02.001

[59] Faleiro JR, El-Shafie HAF, Al-Abdulmohsin AM. Invasion of the red palm weevil, *Rhynchophorus ferrugineus*, in the Gulf and Mediterranean Basin: Current scenarios and control strategies. EPPO Bulletin. 2020;**50**(2):295-308. DOI: 10.1111/epp.12620

[60] Dembilio Ó, Jacas JA. Biology and management of red palm weevil. In: Reddy GK, Davis AL, editors. Recent Advances in Entomological Research: Insect Physiology, Behaviour and Ecology. Cham: Springer; 2018. pp. 13-36. DOI: 10.1007/978-3-319-74059-5_2

[61] Hoddle MS, Antony B, El-Shafie HA, Chamorro ML, Milosavljević I, Löhr B, et al. Taxonomy, biology, symbionts, omics, and management of rhynchophorus palm weevils (Coleoptera: Curculionidae: Dryophthorinae). Annual Review of Entomology. 2024;**69**(1):455-479

[62] Francardi V, Benvenuti C, Roversi PF, Rumine P, Barzanti G. Entomopathogenicity of *Beauveria bassiana* (Bals.) Vuill. And *Metarhizium anisopliae* (Metsch.) Sorokin isolated from different sources in the control of *Rhynchophorus ferrugineus* (Olivier) (Coleoptera Curculionidae). Redia. 2012;**95**:49-55

[63] Hoddle MS, Al-Abbad AH, El-Shafie HAF, Faleiro JR, Sallam AA, et al. Assessing the impact of area-wide pheromone trapping, pesticide applications, and eradication of infested date palms for *Rhynchophorus ferrugineus* (Coleoptera: Curculionidae) management in Al Ghowaybah, Saudi Arabia. Crop Protection. 2013;**53**:152-160

[64] Rani L, Thapa K, Kanojia N, Sharma N, Singh S, Grewal AS, et al. An extensive review on the consequences of chemical pesticides on human health and environment. Journal of Cleaner Production. 2021;**283**:124657

[65] Ali S, Ullah MI, Sajjad A, Shakeel Q, Hussain A. Environmental and health effects of pesticide residues. In: Sustainable Agriculture Reviews 48: Pesticide Occurrence, Analysis and Remediation Volume 2 Analysis. Berlin/Heidelberg, Germany: Springer; 2021. pp. 311-336

[66] Rezende-Teixeira P, Dusi RG, Jimenez PC, Espindola LS, Costa-Lotufo LV. What can we learn from commercial insecticides? Efficacy, toxicity, environmental impacts, and future developments. Environmental Pollution. 2022;**300**:118983. DOI: 10.1016/j.envpol.2022.118983

[67] Chakravarthy AK, Thyagaraj NE. The palm squirrel in coconut plantations: Ecosystem services by therophily. Mammalia. 2012;**76**:2. DOI: 10.1515/mammalia-2011-0073

[68] Wakil W, Yasin M, Qayyum MA, Ghazanfar MU, Al-Sadi AM, Bedford G, et al. Resistance to commonly used insecticides and phosphine fumigant in red palm weevil, Rhynchophorusferrugineus (Olivier) in Pakistan. PLoS One. 2018;**13**(7):e0192628. DOI: 10.1371/journal.pone.0192628

[69] Nelson PN, Sheaves M, Cécile Bessou CI, Lénaïc Pardon CI, Lim HS, Kookana RS. Modelling environmental impacts of agriculture, focusing on oil palm. In: Achieving Sustainable Cultivation of Oil Palm. Vol. 2. Cambridge, UK: Burleigh Dodds Science Publishing; 2018. pp. 285-334. DOI: 10.19103/AS.2017.0018.24

[70] Zaller JG, Zaller JG. Pesticide impacts on the environment and humans. In: Daily Poison: Pesticides-an Underestimated Danger. Cham, Switzerland: Springer Nature; 2020. pp. 127-221

[71] Gullino ML, Albajes R, Al-Jboory I, Angelotti F, Chakraborty S, Garrett KA, et al. Climate change and pathways used by pests as challenges to plant health in agriculture and forestry. Sustainability. 2022;**14**:12421

[72] Siddiqui JA, Fan R, Naz H, Bamisile BS, Hafeez M, Ghani MI, et al. Insights into insecticide-resistance mechanisms in invasive species: Challenges and control strategies. Frontiers in Physiology. 2023;**13**:1112278. DOI: 10.3389/fphys.2022.1112278

[73] Sponsler DB, Grozinger CM, Hitaj C, Rundlöf M, Botías C, Code A, et al. Pesticides and pollinators: A socioecological synthesis. Science of the Total Environment. 2019;**662**:1012-1027. DOI: 10.1016/j.scitotenv.2019.01.016.73

[74] Janssen A, van Rijn PC. Pesticides do not significantly reduce arthropod pest densities in the presence of natural enemies. Ecology Letters. 2021;**24**(9):2010-2024

[75] Bale JS, Van Lenteren JC, Bigler F. Biological control and sustainable food production. Philosophical Transactions of the Royal Society B: Biological Sciences. 2008;**363**(1492):761-776

[76] van Lenteren JC, Bolckmans K, Köhl J, Ravensberg WJ, Urbaneja A. Biological control using invertebrates and microorganisms: Plenty of new opportunities. BioControl. 2018;**63**:39-59

[77] Butt TM, Goettel MS. Bioassays of entomogenous fungi. In: Navon A, Ascher KRS, editors. Bioassays of Entomopathogenic Microbes and Nematodes. Wallingford, UK: CAB International; 2000. pp. 141-195. DOI: 10.1079/9780851994222.0141

[78] Quesada-Moraga E, Santos-Quiros R, Valverde-Garcia P, Santiago-Álvarez C. Virulence, horizontal transmission, and sublethal reproductive effects of *Metarhizium anisopliae* (anamorphic fungi) on the German cockroach (Blattodea: Blattellidae). Journal of Invertebrate Pathology. 2004;**87**:51-58

[79] Ment D, Levy N, Allouche A, Davidovitz M, Yaacobi G. Efficacy of Entomopathogenic fungi as prevention against early life stages of the red palm weevil, *Rhynchophorus ferrugineus* (Coleoptera: Curculionidae) in laboratory and greenhouse trials. Insects. 2023;**14**(12):918

[80] Sutanto KD, Al-Shahwan IM, Husain M, Rasool KG, Mankin RW, Aldawood AS. Field evaluation of promising indigenous entomopathogenic fungal isolates against red palm weevil, *Rhynchophorus ferrugineus* (Coleoptera: Dryophthoridae). Journal of Fungi. 2023;**9**(1):68

[81] Abdel-Rahman IE, Abdel-Raheem MA. Using entomopathogenic fungi as bio agents control on the red palm weevil, *Rhynchophorus ferrugineus* (Olivier) (Coleoptera: Curculionidae). Mortality. 2018;**13**(22):32

[82] Wakil W, Yasin M, Muhammad S-I. Effects of single and combined

applications of entomopathogenic fungi and nematodes against *Rhynchophorus ferrugineus* (Olivier). Scientific Reports. 2017;7(1):5971

[83] Gindin G, Levski S, Glazer I, Soroker V. Evaluation of the entomopathogenic fungi *Metarhizium anisopliae* and *Beauveria bassiana* against the red palm weevil *Rhynchophorus ferrugineus*. Phytoparasitica. 2006;**34**:370-379

[84] El-Sufty R, Al-Awash SA, Bgham AL, S, Shahdad AS, AL Bathra AH. Pathogenicity of the fungus *Beauveria bassiana* (Bals.) Vuill to the red palm weevil *Rhynchophorus ferrugineus* (Oliv.) (Col.: Curculionidae) under laboratory and field conditions. Egyptian journal of biological. Pest Control. 2009;**19**(1):81-85

[85] Guerri-Agullo B, Lopez-Follana R, Asensio L, Barranco P, Lopez-Llorca L. Use of a solid formulation of *Beauveria bassiana* for biocontrol of the red palm weevil *Rhynchophorus ferrugineus* (Col., Dryophoridae) under field conditions in Spain. Florida Entomologist. 2011;**94**:737-747

[86] Gangwar P, Trivedi M, Tiwari RK. Entomopathogenic bacteria. In: Omkar, editor. Microbial Approaches for Insect Pest Management. Singapore: Springer; 2021. DOI: 10.1007/978-981-16-3595-3_2

[87] Dara SK. Entomopathogenic Microorganisms: Modes of Action and Role in IPM. Agriculture and Natural Blogs. eJournal on Production and Pest Management Practices for Strawberries and Vegetables. University of California; 2017. 7 p

[88] Bravo A, Gill SS, Soberón M. Mode of action of bacillus thuringiensis cry and Cyt toxins and their potential for insect control. Toxicon. 2007;**49**(4):423-435. DOI: 10.1016/j.toxicon.2006.11.022

[89] Pu YC, Ma TL, Hou YM, Sun M. An entomopathogenic bacterium strain bacillus thuringiensis as a biological control agent against the red palm weevil *Rhynchophorus ferrugineus* (Col., Curculionidae). Pest Management Science. 2017;**73**:1494-1502

[90] Salama HS, Foda MS, El-Bendary MA, Abdel-Razek A. Infection of red palm weevil, *Rhynchophorus ferrugineus*, by spore-forming bacilli indigenous to its natural habitat. Egyptian Journal of Pest Science. 2004;**77**:27-31

[91] Dolinski C, Lacey LA. Microbial control of arthropod pests of tropical tree fruits. Neotropical Entomology. 2007;**36**:161-179

[92] Lacey LA, Shapiro-Ilan DI. Microbial control of insect pests in temperate orchard systems: Potential for incorporation into IPM. Annual Review of Entomology. 2008;**53**:121-144

[93] Grewal PS, Lewis EE, Gaugler R. Response of infective stage parasites (Nematoda: Steinernematidae) to volatile cues from infected hosts. Journal of Chemical Ecology. 1997;**23**:503-515

[94] Senthil-Nathan S. A review of biopesticides and their mode of action against insect pests. In: Environmental Sustainability: Role of Green Technologies. New Delhi: Springer; 2014. pp. 49-63

[95] Manzoor M, Ahmad JN, Sharif MZ, Majeed D, Kiran H, Jafir M, et al. Comparative effectiveness of entomopathogenic nematodes against red palm weevil (*Rhynchophorus ferrugineus*) in Pakistan. Journal of Entomology and Zoology Studies. 2017;**5**:756-760

[96] Gozel U, Gozel C, Yurt C, Inci D. Efficacy of entomopathogenic nematodes

on the red palm weevil *Rhynchophorus ferrugineus* (Olivier, 1790) (Col., Curculionidae) larvae. International Journal of Bioassays. 2015;**4**:4436-4439

[97] Saleh MME, Alheji M. Biological control of red palm weevil with entomopathogenic nematodes in the eastern province of Saudi Arabia. Egyptian Journal of Biological Pest Control. 2023;**13**(1/2):55-59

[98] Ibrahim M, Loulou A, Brouk A, et al. Parasites rather than phoronts: Teratorhabditis synpapillata nematodes reduce lifespan of their Rhynchophorus ferrugineus host in a life stage-dependent manner. Ecology and Evolution. 2021;**11**(18):12596-12604

[99] Loulou A, Fakhraoui A, M'rabet-Samaali B. Interactions between microorganisms associated with red palm weevil, *Rhynchophorus ferrugineus*, in Tunisia. Euro-Mediterranean Journal of Environmental Integration. 2023;**8**:493-505. DOI: 10.1007/s41207-023-00382-2

[100] Abraham VA, Kurian C. Chelisoches moris F.(Forficulidae: Dermaptera), A Predator on Eggs and Early Instar Grubs of the Red Palm Weevil Rhynchophorus ferrugineus F.(Curculionidae: Coleoptera). 1973. pp. 147-152

[101] Mazza G, Francardi V, Simoni S, Benvenuti C, Cervo R, Faleiro JR, et al. An overview on the natural enemies of Rhynchophorus palm weevils, with focus on *R. Ferrugineus*. Biological Control. 2014;77:83-92

[102] Sarwar M. Biological control of red palm weevil *Rhynchophorus ferrugineus* (Coleoptera: Curculionidae) by the natural enemies. International Journal for Research in Biology & Pharmacy. 2016;**2**(7):22-35. DOI: 10.53555/bp.v2i7.131

[103] Iyer CSV. Two interesting and unrecorded enemies of the palm beetle, *Rhynchophorus ferrugineus*. Indian Journal of Entomology. 1940;**2**:98

[104] Murphy ST, Briscoe BR. The red palm weevil as an alien invasive: Biology and the prospects for biological control as a component of IPM. Biocontrol News and Information. 1999;**20**(1):35-46

[105] Moura JIL, Mariau D, Delabie JHC. Efficacy of Paratheresia menezesi Townsend (Diptera: Tachinidae) for natural biological control of *Rhynchophorus palmarum* L. (Coleoptera: Curculionidae). Oleagineux. 1993;**48**:219-223

[106] Nirula KK. Investigations on the pests of coconut palm. Part. *Rhynchophorus ferrugineus*. Indian Coconut Journal. 1956;**9**:229-247

[107] Peter C. A note on the mites associated with *Rhynchophorus ferrugineus* in Tamil Nadu. Journal of Insect Science. 1989;**2**:160-161

[108] Dembilio Ó, Jaques JA. Bio-ecology and integrated management of the red palm weevil, *Rhynchophorus ferrugineus* (Coleoptera: Curculionidae), in the region of Valencia (Spain). Hellenic Plant Protection Journal. 2012;**5**:1-12

[109] Abdullah MAR. Toxicological and histopathological studies of Boxus chinensis oil and precocene II on larvae of the red palm weevil, *Rhynchophorus ferrugineus* (Oliver) (Coleoptera: Curculionidae). Egyptian Academic Journal of Biological Sciences. 2009;**2**(2):45-54

[110] Al-Deeb MA, Bin Muzaffar S, Mohammad SE. Interactions between Phoretic mites and the Arabian rhinoceros beetle, Oryctes agamemnon arabicus. Journal of

Insect Science. 2012;**12**(1):128. DOI: 10.1673/031.012.12801

[111] Slimane-Kharrat S, Ouali O. Mites associated with the red palm weevil (*Rhynchophorus ferrugineus*) in Tunisia. Tunisian Journal of Plant Protection. 2019;**14**(2):29-38

[112] Sarwar M, Ashfaq M, Ahmed A, Randhawa MAM. Assessing the potential of assorted plant powders on survival of Caloglyphus grain mite (Acari:Acaridae) in wheat grain. Agricultural and Food Sciences. 2013;**13**(2):106

[113] Shannag HK, Capinera JL, Freihat NM. Effects of neembased insecticides on consumption and utilization of food in larvae of *Spodoptera eridania* (Lep., Noctuidae). Journal of Insect Science. 2015;**15**:152-159

[114] Elrehawy ES, ElDoksch HA. The insecticidal and antifeedant activity of *Calotropis procera* latex and foliar extracts against the red palm weevil, *Rhynchophorus ferrugineus* (Olivier) (Coleoptera: Curculionidae). Egyptian Academic Journal of Biological Sciences, F. Toxicology & Pest Control. 2022;**14**(2):25-33

[115] Ahmed FA, Hussein KT, Gad MI. Biological activity of four plant oils, against the red palm weevil, *Rhynchophorus ferrugineus* (Oliver), (Coleoptera: Curculionidae). Journal of Bioscience and Applied Research. 2015;**1**(5):213-222

[116] Al-Dawsari MM. Insecticidal potential of cardamom and clove extracts on adult red palm weevil *Rhynchophorus ferrugineus*. Saudi Journal of Biological Sciences. 2019;**27**(1):195-201

[117] Bajwa WI, Kogan M. Compendium of IPM Definitions (CID)-What Is IPM and How is it Defined in the Worldwide Literature? Corvallis: Integrated Plant Protection Center/Oregon State University; IPPC Publication no. 998; 2022

[118] Zaluck MP, Adamson D, Furlong MJ. The future of IPM: Whither or wither? Australian Journal of Entomology. 2009;**48**(2):85-96. DOI: 10.1111/j.1440-6055.2009.00690.x

[119] Anderson JA, Ellsworth PC, Faria JC, Head GP, Owen MDK, Pilcher CD, et al. Genetically engineered crops: Importance of diversified integrated pest. Frontiers in Bioengineering and Biotechnology. 2019;7:24

[120] War AR, Paulraj MG, Ahmad T, Buhroo AA, Hussain B, Ignacimuthu S, et al. Mechanisms of plant defense against insect herbivores. Plant Signaling & Behavior. 2012;**7**(10):1306-1320

[121] Ju RT, Wang F, Wan FH, Li B. Effect of host plants on development and reproduction of *Rhynchophorus ferrugineus* (Olivier) (Coleoptera: Curculionidae). Journal of Pest Science. 2011;**84**:33-39

[122] Yasin M, Wakil W, El-Shafie HAF, et al. Potential role of microbial pathogens in control of red palm weevil (Rhynchophorus ferrugineus)-A review. Entomological Research. 2017;**47**(4):219-234

[123] Harrison RL, Rowley DL, Funk CJ. The complete genome sequence of Plodia interpunctella granulovirus: Evidence for horizontal gene transfer and discovery of an unusual inhibitor-of-apoptosis gene. PLoS One. 2016;**11**(7):e016038995

[124] Gopinadhan PB, Mohandas N, Nair KPV. Cytoplasmic polyhedrosis virus infecting red palm weevil

of coconut. Current Sciences. 1990;**59**:577-580

[125] El-Minshawy AM, Hendi RA, Gadelhak GG. Viability of stored polyhedrosis virus of the red palm weevil *Rhynchophorus ferrugineus* (Olivier) (Col., Curculionidae). In: FAO/IAEA International Conference on Area-Wide Control of Insect Pests: Integrating the Sterile Insect and Related Nuclear and Other Techniques (IAEA-CN-131). Vienna, Austria: International Atomic Energy Agency; 2005

[126] Banerjee A, Dangar TK. Pseudomonas aeruginosa, a facultative pathogen of red palm weevil, *Rhynchophorus ferrugineus*. World Journal of Microbiology & Biotechnology. 1995;**11**:618-620. DOI: 10.1007/BF00361002

[127] Sarwar M, Shad NA, Batool R. Entomopathogenic protozoa roles in the management of insects pest populations. In: Biopesticides in Organic Farming. CRC Press; 2021. pp. 129-136

[128] Ragaei M, Gesraha MA, Mohamed H, El-Shishtawi R. Symbiotic flagellated protozoa isolated from red palm weevil, *Rhynchophorus ferrugineus* (Olivier), (Coleoptera: Curculionidae). Australian Journal of Basic and Applied Sciences. 2009;**3**(2):604-606

[129] De Clercq P, Mason PG, Babendreier D. Benefits and risks of exotic biological control agents. BioControl. 2011;**56**:681-698

[130] Barah P, Bones AM. Multidimensional approaches for studying plant defence against insects: From ecology to omics and synthetic biology. Journal of Experimental Botany. 2015;**66**:479-493

[131] Dias GB, Altammami MA, El-Shafie HAF, Alhoshani FM, Al-Fageeh MB, Bergman CM, et al. Haplotype-resolved genome assembly enables gene discovery in the red palm weevil *Rhynchophorus ferrugineus*. Scientific Reports. 2021;**11**(1):9987. DOI: 10.1038/s41598-021-89091-w

[132] Alphey N, Bonsall MB. Genetics-based methods for agricultural insect pest management. Agricultural and Forest Entomology. 2018;**20**:131-140

[133] Sethuraman A, Janzen FJ, Weisrock DW, Obrycki JJ. Insights from population genomics to enhance and sustain biological control of insect pests. Insects. 2020;**11**:462

[134] De Souza PI, Doss AA, Vindiola BG, Brown DJ, Ettinger CL, Stajich JE, et al. Efficient CRISPR/Cas9-mediated genome modification of the glassy-winged sharpshooter *Homalodis cavitripennis* (Germar). Scientific Reports. 2022;**12**:6428

[135] Manee MM, Alqahtani FH, Al-Shomrani BM, El-Shafie HA, Dias GB. Omics in the red palm weevil *Rhynchophorus ferrugineus* (Olivier) (Coleoptera: Curculionidae): A bridge to the pest. Insects. 2023;**14**:255

[136] Wang L, Zhang X-W, Pan L-L, Liu W-F, Wang D-P, et al. A large-scale gene discovery for the red palm weevil *Rhynchophorus ferrugineus* (Coleoptera: Curculionidae). Insect Sciences. 2013;**20**:689-702

[137] Yin A, Pan L, Zhang X, Wang L, Yin Y, Jia S, et al. Transcriptomic study of the red palm weevil *Rhynchophorus ferrugineus* embryogenesis. Insect Science. 2015;**22**(1):65-82

[138] Hazzouri KM, Sudalaimuthuasari N, Kundu B, Nelson D, Al-Deeb MA, Le Mansour A, et al. The genome of pest *Rhynchophorus ferrugineus* reveals gene

families important at the plant-beetle interface. Communications Biology. 2020;**3**(1):323

[139] Sudalaimuthuasari N, Kundu B, Hazzouri KM, Amiri KMA. Near-chromosomal-level genome of the red palm weevil (*Rhynchophorus ferrugineus*), a potential resource for genome-based pest control. Scientific data. 2024;**11**:45. DOI: 10.1038/s41597-024-02910-3

[140] Manee MM. Draft genome assembly of adult female red palm weevil *Rhynchophorus ferrugineus*. Zenodo. 2022;**11**(1):45. DOI: 10.5281/zenodo.6878576

[141] Antony B, Soffan A, Jakše J, Abdelazim MM, Aldosari SA, Aldawood AS, et al. Identification of the genes involved in odorant reception and detection in the palm weevil *Rhynchophorus ferrugineus*, an important quarantine pest, by antennal transcriptome analysis. BMC Genomics. 2016;**17**(1):1-22. DOI: 10.1186/s12864-016-2362-6

[142] Gonzalez F, Johny J, Walker WB III, Guan Q, Mfarrej S, Jakše J, et al. Antennal transcriptome sequencing and identification of candidate chemoreceptor proteins from an invasive pest, the American palm weevil, *Rhynchophorus palmarum*. Scientific Reports. 2021;**11**(1):8334

[143] Soffan A, Antony B, Abdelazim M, Shukla P, Witjaksono W, Aldosari SA, et al. Silencing the olfactory co-receptor RferOrco reduces the response to pheromones in the red palm weevil, *Rhynchophorus ferrugineus*. PLoS One. 2016, 2016;**11**(9):1-19. DOI: 10.1371/journal.pone.0162203

[144] Harith-Fadzilah N, Lam SD, Haris-Hussain M, Ghani IA, Zainal Z, Jalinas J, et al. Proteomics and interspecies interaction analysis revealed abscisic acid signalling to be the primary driver for oil palm's response against red palm weevil infestation. Plants. 2021;**10**(12):1-26. DOI: 10.3390/plants10122574

[145] Flowers JM, Hazzouri KM, Lemansour A, Capote T, Gros-Balthazard M, Ferrand S, et al. Patterns of volatile diversity yield insights into the genetics and biochemistry of the date palm fruit Volatilome. Frontiere in Plant Science. 2022;**13**:1-14. DOI: 10.3389/fpls.2022.853651

[146] Rasool KG, Khan MA, Aldawood AS, Tufail M, Mukhtar M, et al. Identification of proteins modulated in the date palm stem infested with red palm weevil (*Rhynchophorus ferrugineus* Oliv.) using two-dimensional differential gel electrophoresis and mass spectrometry. International Journal of Molecular Science. 2015;**16**:19326-19346

[147] Abdel-Banat B, El-Shafie HA. Genomics approaches for insect control and insecticide resistance development in date palm. The Date Palm Genome. 2021;**2**:215-248

[148] Alshammari N, Alazmi M, Alanazi NA, et al. A comparative study on the microbial communities of *Rhynchophorus ferrugineus* (red palm weevil)-infected and healthy palm trees. Arab Journal of Scientific Engineer. 2022;**47**:6735-6746. DOI: 10.1007/s13369-021-05979-9

[149] Knipling EF. Possibilities of insect control or eradication through the use of sexually sterile males. Journal of Economic Entomology. 1 Aug 1955;**48**(4):459-462. DOI: 10.1093/jee/48.4.459

[150] Klassen W. Area-wide integrated pest management and the sterile insect technique. In: Dyck VA, Hendrichs J,

Robinson AS, editors. Sterile Insect Technique: Principles and Practice in Area-Wide Integrated Pest Management. Dordrecht, The Netherlands: Springer; 2005. pp. 39-68. ISBN 978-1-4020-4051-1

[151] Cristofaro M, Fornari C, Mariani F, Cemmi A, Guedj M, Ben Jamaa ML, et al. Effects of γ-irradiation on mating behavior of red palm weevil, *Rhynchophorus ferrugineus* (Olivier, 1790) (Coleoptera: Dryophthoridae). Insects. 2023;**14**(7):661

[152] Rahalkar GW, Harwalkar MR, Rananavare HD, Shantaram K, Ayengar ARG. Laboratory studies on radiation sterilization of the red palm weevil (*Rhynchophorus ferrugineus* Oliv.) males. Journal of Plant Crops. 1973;**1**:141-145

[153] Krishnakumar R, Maheswari P. Assessment of the sterile insect technique to manage red palm weevil *Rhynchophorus ferrugineus* in coconut. In: Vreysen MJB, Robinson AS, Hendrichs J, editors. Area-Wide Control of Insect Pests. Dordrecht, The Netherlands: Springer; 2007. pp. 475-485. ISBN 978-1-4020-6058-8

[154] Prabhu ST, Dongre TK, Patil RS. Effect of irradiation on the biological activities of red palm weevil, *Rhynchophorus ferrugineus* Olivier. Karnataka. Journal of Agricultural Science. 2010;**23**:186-188

[155] Whitten M, Mahon R. Misconceptions and constraints. In: Dyck VA, Hendrichs J, Robinson AS, editors. Sterile Insect Technique: Principles and Practice in Area-Wide Integrated Pest Management. Dordrecht, The Netherlands: Springer; 2005. pp. 601-626. ISBN 978-1-4020-4051-1

[156] Belvedere S, Arnone S, Cristofaro M, La Marca A, De Biase A. Paternity analyses for the planning of SIT projects against the red palm weevil. Insects. 2023;**14**(4):326. DOI: 10.3390/insects14040326

[157] Al-Ayedh HY, Rasool KG. Determination of the optimum sterilizing radiation dose for control of the red date palm weevil *Rhynchophorus ferrugineus* Oliv. (Coleoptera: Curculionidae). Crop Protection. 2010;**29**(12):1377-1380

[158] Mazza G, Inghilesi AF, Stasolla G, Cini A, Cervo R, et al. Sterile *Rhynchophorus ferrugineus* males efficiently impair reproduction while maintaining their sexual competitiveness in a social context. Journal of Pest Science. 2016;**2016**(89):459-468

[159] El-Ezaby FA. Injection as a method to control the red Indian date palm weevil Rhynchophorus ferrugineus. Arab Journal Plant Protection. 1997;**15**:31-38

[160] Ben-Amor R, Aguayo E, de Miguel-Gomez MD. The competitive advantage of the Tunisian palm date sector in the Mediterranean region. Spanish Journal of Agricultural Research. 2015;**13**:e0101

[161] Ferry M, Cousin R, Chabernaud D, Ferrero F. An effective strategy to obtain very rapidly the red palm weevil decline in an area planted with ornamental palms. Journal of Arab Society for Plant Protection. 2019;**37**(2):188

[162] Koubaa A, Aldawood A, Saeed B, Hadid A, Ahmed M, Saad A, et al. Smart palm: An IoT framework for red palm weevil early detection. Agronomy. 2020;**10**:987. DOI: 10.3390/agronomy10070987

[163] de Groot RS, Matthew WA, Roelof MJA. Typology for the classification, description and valuation of ecosystem functions, goods and services. Ecological Economics.

2002;**41**(3):393-408. DOI: 10.1016/S0921-8009(02)00089-7

[164] Laucht S, Brulez K, Hanisch J, Blakey A, Weyman G, Ludwigs JD, et al. Ornamental Phoenix palm trees as habitat for fauna in the Mediterranean region–results from a full year monitoring. Biodiversity Data Journal. 2024;**12**:e123144

[165] Brockerhoff EG, Andrew ML, Herve J. The ecology offorest insect invasions and advances in their management. Canadian Journal of Forest Research. 2006;**36**:263-268

[166] Fortuna TM, Le Gall P, Mezdour S, Calatayud PA. Impact of invasive insects on native insect communities. Current Opinion in Insect Science. 2022;**51**:100904

[167] Pejchar L, Mooney HA. Invasive species, ecosystem services and human well-being. Trends in Ecology and Evolution. 2009;**24**:497-504. DOI: 10.1016/j.tree.2009.03.016

[168] Vilà M, Espinar JL, Hejda M, Hulme PE, Jarošík V, Maron JL, et al. Ecological impacts of invasive alien plants: A meta-analysis of their effects on species, communities and ecosystems. Ecology Letters. 2011;**14**:702-708

[169] Bessou C, Rival A, Levang P, Feintrenie L, Bosc PM, Cheyns E, et al. Sustainable palm oil production project synthesis: Understanding and anticipating global challenges. CIFOR. 2017;**165**:1-6

[170] Food and Agriculture Organization (FAO). Red Palm Weevil: Guidelines on Management Practices. Rome: FAO; 2020. Available from: https://www.fao.org/documents/card/en/c/ca7703en

[171] Hussain A, Rizwan-Ul-Haq M, Al-Jabr A, Al-Ayied H. Managing invasive Popula-Tions of red palm weevil: A worldwide perspective. Journal of Food Agriculture and Environment. 2013;**11**:456-463

[172] Ahmad I. Integrated pest management of *Rhynchophorus ferrugineus* Olivier: An efficient approach to reduce infestation in date palm trees. Pakistan Journal of Zoology. 2022;**54**:927

[173] El-Shafie HAF. Integrated insect pest management [Internet]. In: Pests Control and Acarology. IntechOpen; 2020. DOI: 10.5772/intechopen.81827

[174] Deguine JP, Aubertot JN, Flor RJ, Lescourret F, Wyckhuys KAG, Ratnadass A. Integrated Pest management: Good intentions, hard realities. A review. Agronomy for Sustainable Development. 2021;**41**:38

[175] Alotaibi BA, Ahmed A, Al-Zaidi AA, Kassem HS. Adoption of integrated pest management for red palm weevil control among farmers in Saudi Arabia. Horticulturae. 2022;**8**:1005

[176] Sallam AA, El-Shafie HAF, Al-Abdan S. Influence of farming practices on infestation by red palm weevil *Rhynchophorus* (Olivier) in date palm: A case study. International Research Journal of Agricultural Science and Soil Science. 2012;**2**:370-376

[177] Faleiro JR, Al-Shawaf AM, Al-Dandan AM, Al-Odhayb A, Al-Rudayni A, Abdallah AB, et al. Controlled release products for managing insect pests. Outlooks on Pest Management. 2016;**27**:175-180

[178] Abdedaiem S, Nasr N, Ferry M. Socioeconomic studies and approaches to involve farmers in the fight against red palm weevil (RPW). In: Presented at the "Scientific Consultation and

High-Level Meeting on Red Palm Weevil Management", Organized by FAO and CIHEAM. Rome: FAO; 2017. pp. 29-31

[179] Aldobai S, Ferry M. Proposed multidisciplinary and multi-regional strategy for the management of red palm weevil. In: Presented at the "Scientific Consultation and High-Level Meeting on Red Palm Weevil Management", Organized by FAO and CIHEAM. FAO; 2017. pp. 29-31

[180] Yu R, Leung P. Optimal pestmanagement: A reproductive pollutant perspective. International Journal of Pest Management. 2006;**52**:155-166

[181] Rigakis I, Potamitis I, Tatlas N-A, Potirakis SM, Ntalampiras S. TreeVibes: Modern tools for global monitoring of trees for borers. Smart Cities. 2021;**4**:271-285

Chapter 6

Metal Up-Take in Biological Systems: A Review

Carlie Stransky Ogilvie, Graeme A. Spiers and Frank F. Mallory

Abstract

This manuscript reviews the rise of environmental awareness since WWII and the resulting scientific and technological developments implemented since then to ensure healthy populations and a sustainable environment. Essential and non-essential metals are discussed in relation to their positive and negative impacts and their relationship to ecotoxicology, anthropogenic impacts, and natural sources. Metals originating from the environment are re-introduced, often in bioavailable forms from anthropogenic activities. Their movement throughout our natural systems is not always visible, making them difficult to study and understand. Although the effects of individual metals has been studied extensively, the amalgamated effects of multiple metals on individuals are not well understood. There are very few studies on the effects of many metals from multiple trophic levels. Understanding this level of ecotoxicology can be useful in understanding the in-situ effects that urban development and industries such as mining and forestry have on the local environment.

Keywords: environment, heavy metals, ecotoxicology, anthropogenic, natural sources, conservation

1. Introduction

As human populations continue to grow, understanding the association between the change in density and development is vital to ensure healthy populations and environmental sustainability. Since the 1900s, especially following the Industrial Revolution and World War II, there has been an awareness that environmental resources are in finite supply and are becoming increasingly degraded [1]. The Industrial Revolution brought with it resource-harvesting industries whose practices quickly became the center of criticism by a rising environmental movement [1–3]. Research brought to light how important maintaining environmental health is for the well-being of planetary life, and since then, significant progress has been made to monitor and solve the immeasurable number of environmental problems that have been created [2]. The rise of environmentalism and conservation efforts was also promoted by publications such as Rachel Carson's Silent Spring [4], which had an enormous impact on how society regarded and treated industrial toxins in the environment [1, 3]. With Carson's publication, issues such as the negative effects of pesticides on non-target species and the development of resistance, chemical persistence, and imbalances were publicized and debated [5]. Research programs around the world blossomed, and chemical screening programs were initiated to look at the toxicity and potential hazards of chemicals to birds and mammals, including humans [6, 7]. As a result, long-term sustainable environmental monitoring programs were initiated [5]. Despite the sudden surge of environmental awareness and conservation activity, by the mid-1990s, the environment continued to degrade, fragment, and be polluted at an alarming rate [2]. However, some significant progress was made to solve and monitor environmental problems [2], as it was recognized that industries such as mining, hydro, oil, and nuclear could cause serious adverse effects on the environment. As a result, it was important that these industries adhere to government regulations, or companies could risk severe fines, the loss of a positive public image, avoidance from potential investors, and being forced to close [8]. In Canada, to mitigate environmental degradation, conservation regulations such as the Species at Risk Act and the Canadian Environmental Protection Act worked to address a variety of issues such as pollution prevention, wildlife protection, and emergency management [9]. For industrial purposes, biomonitoring, which involves measuring the burden of toxic elements or compounds, can be an effective method to detect and assess hazardous contaminants and chemicals in an environment [2]. As industries continued to expand in both size and quantity, so did the need to ensure that their practices were not harming the local environment or the species that depend on it [10, 11]. At the time, one of the main groups of pollutants responsible for the degradation of ecosystem health was metals. Metals, due to their toxicity, bio-accumulative, and non-biodegradable nature, can be incredibly dangerous to aquatic, terrestrial, and human life systems [12]. Once accumulated in animals, plants, or humans, metals can be difficult to detoxify, metabolize, or excrete [13]. Ecotoxicology studies the movements of environmental contaminants through the environment and their effects on plants, animals, and ecosystems [14]. The concern of poisoning wildlife initially began in the nineteenth century and at first, focused primarily on identifying environmental problems rather than contaminants in tissues [5]. However, as science progressed, reports on pheasant (*Phasianus colchicus*) and waterfowl mortality caused by lead ingestion began to appear in literature [15–17]. To understand the movement of contaminants within organisms and through food chains as well as injuries to both individuals and communities, examining tissue residues of these contaminants is key to ecotoxicology [14]. The logic of using tissues in wildlife toxicology was introduced by Bill and Lucille Stickel [18], who described a method of using bird tissues to investigate organochlorine pesticide poisoning [14]. However, as early as 1919, both qualitative and quantitative determination were used to describe the presence of lead in the caeca and stomach of waterfowl [19, 20]. Although soil and sediment analyses are important, tissue contaminant is much more important for toxicologists as it can give more insight into global distribution of a contaminant or even explain the reason and origin of mass die-offs [14].

In summary, this chapter is a review of original research related to heavy metals in the environment and the conservation measures and impacts they have, as some are essential and others detrimental. The extensive reference section allows researchers to access the original research that was done and to access the numerous methodologies that were used.

2. What are heavy metals?

The term 'heavy metals' is used to describe a metallic element with a density of more than 3.5–5 g/cm^3 and an atomic number greater than 20 [21, 22]. Their presence can have deleterious effects because they do not degrade and they bio-accumulate in living tissues giving a rise to symptoms of toxicity. Additionally, there are a large number of poorly studied metals and metalloids [14]. These compounds and elements can be found in pharmaceuticals, industry effluents, personal care products, pesticides, and much more. Metals occur in two categories: elements that are necessary for growth and organism life but are toxic at high concentrations and elements that are toxic even at low concentrations [22]. At elevated levels, both essential and non-essential metals can negatively affect an organism's development, physiology, pathology, feeding, and reproduction, while increases in radicals can severely damage DNA [23, 24]. Aside from the levels and bioavailability of metals in the area, health risks for mammals depend on many confounding variables such as diet, nutritional status, age, and metabolism [25]. The main sources of these contaminants are anthropogenic: sewage, industrial effluents, fuel combustion, mining, and smelting [22]. Therefore, there is a significant need for industry and government to understand and monitor the causal relationships and manage risks to wildlife from heavy metal-bearing waste solutions and tailings [25]. Metal pollution, although less visible, is a direct and significant polluter of aquatic ecosystems as they are deposited, assimilated, and incorporated in water, binding to sediment particles and aquatic species [26]. The reason for metal concern is primarily due to the phenomena of bioaccumulation. This term describes the process of how metals accumulate in species and to other trophic levels, once they have become bioavailable and enter the food 10 system [27]. Biomagnification is another major concern, as it is the process of contaminants increasing in concentration as they move up trophic levels [28], while accumulating more metals through predation of contaminated prey over time [27]. The primary exposure pathway of such pollutants is ingestion [29]. More specifically, the intake of metals is not only related to the feeding habits of a species, but also the specifics of behavior are important [30]. For example, some of the highest metal levels can be found in insectivores as they feed on earthworms, which have strong metal accumulation capabilities [30]. With such complexities in mind, a multi-trophic level approach, as performed by Saunders [24], is recommended to determine concentrations and bioavailability in the environment. The transfer of metals in food chains is described as bio-concentration factors (BCFs) and bioaccumulation factors (BAFs), and they are based on the knowledge of the linear movement of metals through soil, to plants, insects, mammals, and other organisms [24]. BCF is defined as the ratio of the metal concentration in an organism to the concentration of the original source, which could be water or soil in its steady state. BCF is used for plants and invertebrates and is expressed in the wet weight of tissue and the dry weight of soil. BAF is the ratio of the metal concentration in the organisms to the concentration of the metal in its food at a steady state [24]. It is used to describe the accumulation in birds and mammals and is expressed on a wet-weight basis [27].

Various studies have emphasized the need for further research on the accumulation and fate of metal contaminants at different trophic levels and waterbody types. Metals moving through aquatic food webs to fish, other piscivorous animals, and humans cause serious environmental and human health concerns. High levels of mercury (Hg) in fish from supposedly pristine lakes have resulted in the adoption of conservative fish consumption advisories globally [31–33]. However, pathways of metal movement through different food webs are not well understood. The understanding of metal movement in freshwater systems has been growing significantly since the 1990s, but there are still large gaps in knowledge as most studies of metals focus on a few taxa, single metals, or movement through a single trophic level [34].

For marine environments, some metals have been studied almost exclusively on their own without evaluating the impacts of multiple metals causing symptoms together [35, 36]. Unfortunately, piscivorous wildlife in freshwater habitats has been found to have higher levels of metal exposure than those in marine habitats, stressing the need for further understanding of freshwater systems [37]. Due to a lack of consistency among studies comparing and understanding the transfer processes of different metals across different systems, understanding metal movements through food webs in freshwater lakes has proved to be challenging. Due to technical shortcomings and a significantly poor understanding of community structure, very little metal movement, particularly in the lower trophic levels in freshwater systems, is understood. Furthermore, despite a large amount of research conducted on heavy metal behavior in laboratories on small mammals, less work has been done on understanding metal movements in natural environments and their impacts on wildlife [38].

3. Metals of interest

There are several metals that have significantly negative impacts on organisms. The eight most common pollutant metals listed by the Environmental Protection Agency are arsenic (As), cadmium (Cd), chromium (Cr), copper (Cu), mercury (Hg), nickel (Ni), lead (Pb), and zinc (Zn) [39]. These metals with the exception of Zn, Cu, and Ni are systemic toxicants and are known to induce extensive organ damage, even at low exposure [40]. These metals have been classified as either "known" or "probable" carcinogens based on their exposure and cancer incidence in humans and animals by the United States Environmental Protection Agency and the International Agency on Cancer Research [40]. However, there are a variety of metals that are found in animal and human diets that are essential as they perform several necessary biological functions [41]. A deficiency in any of the required trace metals can result in a variety of deficiency syndromes and diseases, while excessive levels of such metals can lead to cellular and tissue damage leading to diseases, and sometimes death [40]. Some metals of concern have no established biological functions and are considered non-essential metals [40]. The following metals are of concern.

4. Essential metals of concern

4.1 Cobalt (Co)

Co is required for a multitude of biochemical and physiological functions including as a constituent of vitamin B12 [39, 40]. Co alloys are often found in engines, heavy machinery, artificial joints, colorants in glass, ceramics, paints, as catalysts, and as paint driers. Naturally, small amounts of Co are found in rocks, soil, plants, and animals. With the exception of a few hyper-accumulator plant species such as *Berkheyda coddii, Alyssum corsicum, Lamiaceae, and Scropular*, most terrestrial plant species do not bio-concentrate Co [42]. In dissolved or ion form, Co is also found in minute amounts in water. It enters air and water from wind-blown dust, seawater spray, volcanic eruptions, forest fires, and surface water runoff, leaching from rainwater on contaminated rocks and soil [43]. Anthropogenic sources originate from vehicle exhaust, industrial activities from mining and processing of Co-containing ores, coal-fired power plants and incinerators, and from Co alloy and chemical production [43]. A dominant dependent of Co bioavailability is a low pH [42]. Co ions both *in vitro* and *in vivo* are genotoxic and carcinogenic in rodents [44]. Co-metal dust has also been linked to increased lung cancer risk in mammals and induces DNA strand breaks in hamster ovary cells [45]. Animal carcinogenicity data has been shown to be sufficient to make such claims, but not enough data is available to prove it is carcinogenic for humans [44]. Because Co is used less industrially than other metals, there are significantly fewer studies on it, especially on vertebrates [42, 44].

4.2 Chromium (Cr)

Cr is an important regulator of glucose metabolism and cholesterol [46]. It is a common metal in leachates from landfill composts, electroplating, leather tanning, cement, mining, dyeing, fertilizer, and photography industries [47]. Because Cr leaches quickly and is organically complexed in these effluents, it is highly bioavailable and mobile in soil [48]. At elevated levels, Cr is associated with abnormal enzyme activity, changes in blood chemistry, a decreased resistance to pathogenic organisms, alterations in behavior, osmoregulatory upset, modifications in population structure, and in plants the inhibition of photosynthesis [47].

4.3 Copper (Cu)

Cu is essential for many oxidative enzymes, the synthesis of hemoglobin, and antioxidant systems [46, 48]. However, before the 1980s, no biological function was known and Cu was believed to only have adverse effects on metabolic processes with its bioaccumulation capacities [38]. It is a constituent of oxidases, which is important for cartilage formation and regulates redox respiration. When elevated levels are in combination with low pH, Cu can be fatal for aquatic organisms such as fish [39]. If an animal's liver is damaged when Cu is released into the bloodstream, it can damage the kidneys by gorging on the tissues through hemoglobin breakdown. This in turn causes hepatic necrosis and a possibly fatal hemolytic event [49]. High exposure levels have been shown to induce anemia, impair thyroid and liver functions, and reduce food intake, growth, and reproduction [50]. Cu is known to deposit in the brain, skin, liver, pancreas, and myocardium and in humans causes health problems such as anorexia, migraines, allergies, childhood hyperactivity, and learning disorders [47]. Since the 1980s, the biological effects of excess Cu and its mobility within food chains have been poorly documented [38].

4.4 Iron (Fe)

Fe, being one of the most abundant metals on earth, is essential to organisms for hemoglobin, cytochromes, catalysis, and peroxides [39]. However, at elevated levels, Fe can be associated with many physiological impairments, affecting neurobehavioral development, impacting blood enzymes, and causing death [51].

4.5 Manganese (Mn)

Mn at lower levels can play an important role in protecting the mitochondria from free radical damage [39]. A study by Hayat et al. [49] showed that when exposed to elevated levels of Mn for 30 days, three species of carp (*Catla catla, Labeorohita, and Cirrhina mrigala)* experienced decreased weight. They also noted the carp absorbed Mn mainly through gills and not from ingestion. This may be the norm for most species of freshwater fish.

4.6 Nickel (Ni)

Ni is essential for normal growth and reproduction. Ni most commonly enters the body via food and water consumption as well as inhalation. This heavy metal is commonly associated with industries such as non-ferrous metal and mineral processing, paint formulation, electroplating, porcelain enameling, Cu and sulfate manufacturing, and steam-electric power plants. Toxic effects have been documented in humans and laboratory mammals for cardiovascular, respiratory, gastrointestinal, musculoskeletal, hepatic, renal, dermal, ocular, immunology, developmental, neurological, and reproductive systems [47]. Elevated rates can also cause carcinogenic effects when consumed [46]. For sensitive aquatic species, ionic nickel has been observed to be lethal at 11–113 lg/L. However, aquatic bacteria and yeasts are tolerant to Ni. Excretion of Ni is an incredibly slow process compared to its assimilation, causing the kidney to be significantly burdened as Ni tends to accumulate in high levels within it over time [50]. Not enough data has been published to support or refute whether organisms from higher trophic levels are more sensitive to Ni exposure than those from lower trophic levels or whether Ni biomagnifies in aquatic and terrestrial environments [52].

4.7 Selenium (Se)

Se is recognized for its biological importance and is increasingly being recognized as a nutraceutical component [39]. Se is essential for cellular antioxidant defenses [39]. Furthermore, Se can act as a growth factor, has anticancer properties, and is needed for normal thyroid homeostasis [53]. Several studies have observed that Se counteracts the effects of Hg [54, 55] and protects animals from both inorganic Hg and MeHg (methylmercury) [56]. According to a study [55], Se seems to detoxify metals in the livers of higher-topic marine animals. In a laboratory study, animals exposed to both Hg and Se had longer life spans, higher growth rates, and higher weight gains compared to those who received Hg without Se. Additionally, they suffered less tissue damage and other toxic effects [57]. According to Whanger [57], the exact mechanisms that Se uses to protect organisms against Hg are unknown. Whangner [58] found that while Se seems to counteract the neurotoxicity of Hg, it also seems to increase the levels of Hg in the brain in a non-toxic form. Furthermore, Haines et al. [57] found an interesting reaction in the liver and brain as MeHg would de-methylate when exposed to Se. Haines et al. [57] also claimed that research has focused more so on saltwater fish, but more work needs to be done on freshwater fish. They point out that freshwater aquatic mammals such as otters and mink should be exposed to higher levels of Hg than terrestrial mammals and therefore require additional study.

4.8 Zinc (Zn)

Zn is an essential mineral for animals and humans required for normal nucleic acid, protein and membrane metabolism, and cell growth and division [47]. Although it is essential, Zn also has toxic environmental factors [39]. In several studies, high exposure levels of Zn have been shown to induce anemia, impair thyroid and liver function, and decrease food intake, growth, and reproduction [50].

5. Non-essential metals of concern

5.1 Arsenic (As)

As is a widespread element that is found associated with many different minerals in rocks and soils [58]. It is responsible for specific growth stimulation, but high levels are usually obtained through inhalation of dust, and consumption of food, water, and soil, which can lead to DNA damage, respiratory, reproductive, and immunity problems, cell damage, cancer, and even death [39, 59]. Most studies were conducted in laboratories so most of the data resides in an artificial setting. However, in nature, As is taken up by vegetation and various invertebrates that naturally occur on abandoned mine sites [58]. Acute poisoning of As has been reported since the 1800s in domestic and wild animals, but data on wildlife poisoning is scarce. For free-living mammals, metabolic detoxification and rapid excretion is possible. However, as a response to long-term exposure at former mine sites, oxidative stress, histological damage, and some biomarker responses have been associated with As contamination in small mammal species [58].

5.2 Mercury (Hg)

Hg has no known biological function [27]. Being directly mobilized by humans for thousands of years through industrial activities, the atmosphere is the foremost transport pathway of Hg emissions, and then land and ocean/aquatic pathways play an active role in redistributing Hg in terrestrial, freshwater, and marine ecosystems [60]. Both temporal and spatial scales of Hg movement from the atmosphere and then to aquatic and terrestrial environments depend on Hg's chemical and physical form. Once released into the atmosphere, the residence time can be from several months to a year. Elemental Hg (Hg(0)) can be moved long distances before it is oxidized and removed by both particle and gas-phase dry deposition or by precipitation [60]. Hg intoxication can cause effects such as impaired motor coordination. Studies on wildlife populations for Hg intoxication may not always yield population-representative results as only individuals who are active and healthy enough to locate bait set by hunters, fishermen, or trappers would be captured. Therefore, the relationship between Hg, wild animals, and age is still not clear [61]. When assessing Hg toxicity in an organ or individual, the total concentration of the contaminant can be misleading because its toxicity can be dependent on its chemical form [14]. The most bioavailable form of Hg is MeHg (methylmercury), which is an extremely toxic compound that easily crosses biological membranes, accumulates in organisms, which are exposed, and can bio-magnify its way from the bottom to the top of trophic levels in both fish and wildlife [62]. MeHg enters the body mainly from digested food. It is quickly absorbed into the intestines, which then enters the red blood cells and circulates throughout the body [63]. For individuals, laboratory studies show that elevated MeHg exposure can cause significant behavioral, physiological, immunological, neurochemical, reproductive, and histological changes [64]. For populations, impacts of MeHg vary with the life history of the species and local variability of exposure [64]. Many studies document the effects of MeHg in vertebrate individuals, but very few studies look at whether present-day environmental levels of MeHg are a danger to wildlife populations [64]. Conditions that enhance methylation in an environment are low alkalinity, low pH, and high organic carbon concentrations [65]. Dominant zones for methylation are warm, shallow lakes that have organic-rich sediment and wetlands with anoxic waters with shores that have soil drying and rewetting locations [65, 66]. Therefore, due to their anaerobic condition, peatlands, for example, have ideal conditions for promoting methylation [67]. Methylated inorganic Hg quickly enters food webs and biomagnifies sometimes to toxic concentrations. Risk-assessment modeling approaches used for wildlife and Hg often work toward the health of wildlife populations as opposed to the health of individuals, and the conservation of wildlife habitats is a major consideration [64]. However, additions of sulfur dramatically increase methylation and once additions of sulfur are stopped, there is a significant decrease in Hg [68]. According to Scheuhammer et al. [37], there is a sufficient amount of evidence to indicate that in North America, a number of species of wild fish, birds, and mammals consume diets with MeHg concentrations that are toxic to individuals [69]. Because MeHg is so bioavailable and easily bio-accumulates in fish, piscivorous wildlife from higher trophic levels such as otter and mink are at significant risk of over-exposure [70].

5.3 Lead (Pb)

No biological function is known for Pb and toxicity is often associated with cognitive impairment when blood levels are above 100 μg/L [27]. According to the Canadian Soil Quality Guidelines for the Protection of Environmental and Human Health, Pb extraction from ore is often associated with Zn, as the two are often found together in nature [71]. Natural sources of Pb are volcanoes, forest fires, and sea salt [71]. Common pathways for mammals include inhalation of wind-blown lead-laden dust and oral intake of water and food [14]. In the case of a developing fetus, Pb can be transferred through the placenta and from maternal blood to the milk, exposing offspring [72, 73]. Once in the body, Pb targets several physiological systems such as the central nervous system, the hematopoietic system, the renal system, the cardiovascular system, and the gastrointestinal system [25, 74]. Consequently, elevated Pb exposure can cause anemia, kidney damage, hypertension, encephalopathy, injury to peripheral nerves, excessive salivation, intestinal colic, vomiting, and behavior issues such as insomnia, uncoordinated movement, loss of appetite, and lassitude [25]. Extremely high exposure levels may also lead to death [25].

5.4 Cadmium (Cd)

Cd is a widespread element throughout the environment in low concentrations [75]. Cd has no biological function [27], but increased levels of Cd are derived from anthropogenic sources such as fertilizer applications, roasting zinc ores, sewage, and mine waste [75]. Factors such as pH, soil type, and particle size influence Cd mobility [75]. Its toxicity may originate from exposure to respiration or ingestion. A study by Hunter and Johnson [38] revealed that Cd in mammals and invertebrates accumulate in organs such as the liver, kidneys, and bones. Excessive accumulation can lead to tissue dysfunction, DNA damage, and mRNA transcriptional changes in mammal and fish species [13, 68]—Hoffmann et al. [48] studied levels of this non-essential metal in wildlife. Cd exposure in wildlife is often a chronic problem due to slow accumulation via the diet [68]. Little evidence supports the accumulation of Cd in the brain, lung, or testis, even in highly contaminated site [38]. Effects on the kidney often involve tubular dysfunction, tubular cell damage, and glomerular dysfunction. In a study on mice, rats, and voles [70], non-lethal levels of Cd in whole body, liver, and kidney were respectively 0.1–0.4, 0.2–1.5, and < 0.1–5.6 mg/kg. Levels from shrews with an insectivorous diet in uncontaminated areas were found in the range of 1.2–4, 2.3–25.4, 4.1–25.7 in the whole body, liver, and kidney, respectively [70]. Higher levels of Cd have been found in animals from higher trophic levels, indicating biomagnification activity, and older animals show higher levels of Cd, demonstrating bioaccumulation over longer lifetimes [68]. A study on the greater white-toothed shrew (*Crocidura russula*) [71] showed that long biological half-lives of Cd in the kidney and liver indicate that tissue concentrations were dependent on age and the dose, and the duration of the exposure. There has also been evidence that there is a positive relationship between body weight, kidney, and liver concentrations within population samples [73].

5.5 Antimony (Sb)

Sb has received little attention in research due to its limited commercial and industrial use but is most commonly obtained through inhalation and skin contact [44]. Sb is used primarily as a flame retardant in rubber, plastics, textiles, paper, pigments, fireworks, and adhesives. Sb has been found to be a carcinogenic and clastogenic, but does not induce gene mutations in vitro. For people working in smelters and in rodents, inhalation of certain Sb compounds has been reported to cause lung tumors [44].

6. Metal behavior

6.1 Distance from sources

The distance a point-source contaminant is from biota can be seen in the concentrations of metals accumulated in an individual. Rose and Parker [74] studied mineral levels in the plumage of ruffed grouse (*Bonasa urnbellus*) and noted a substantial temporal increase in Cu, Ni, and Fe in the plumage of birds living near Falconbridge, an area within the Greater City of Sudbury. In contrast, birds studied from a site 100 km away from the smelter failed to exhibit any significant change in metal levels over the course of the year [76]. Lepage and Parker [76] confirm the geological and geographical influence of Sudbury, Ontario Ni Cu smelters on local vertebrate populations. However, according to Scheuhammer et al. [37], environments remote from point-source releases of metals such as Hg can contain fish with metal accumulation. Scheuhammer et al. [37] also cautioned that the removal of point-source effluents does not mean the issue of metal contamination has been eliminated. Fish and piscivorous wildlife have been shown to experience elevated exposure long after the inputs have ceased. The size of the home range of an animal can have an impact on the levels of metals individuals are exposed to. A study by Hoffman et al. [48] assessed the levels of Cd, Cu, Fe, and Zn in the kidneys of gray wolves (*Canis lupus*) and a population from Alaska showed a bimodal distribution of metal concentrations. One portion of the population had extremely high levels of all four metals, while levels in the other portion of wolves had significantly lower levels. However, there was no observable pattern related to location, sex, or age to explain the differences. It was determined that the range and dispersal patterns of the wolves affected their metal levels. The wolves with larger ranges appeared to have been in previously contaminated locations, whereas those with smaller ranges were in cleaner areas and had lower levels [48]. It was concluded that to determine metal bioavailability and levels, it is vital to select animals with small ranges to ensure their levels are from local exposures and not from other areas outside their range.

6.2 Trophic level behavior

It is important to note that for most metals, population impacts vary by life history of the species and differences in exposure levels [37]. A study by Chen et al. [77] looked at aqueous metal levels in lakes and how they move through the food web. Chen et al. [77] encourage the investigation of upper and lower trophic levels separately to fully understand metal transfer pathways in aquatic food webs. Their primary finding when studying different trophic levels and metal accumulation in aquatic environments was that water quality is not an effective predictor for metal levels in macro-zooplankton. However, metal in food was a good predictor. Hunter and Johnson [38] also found that for metals such as Cu and Cd, concentrations in surface soils and vegetation were not reliable indicators of food web accumulation between prey and predators, shedding light on the contrasting mobility among different biological systems. Chen et al. [77] showed that trophic relationships varied among different metals, such as Hg, Zn, and Cd bio-magnify, while others such as As and Pb bio-diminish. This study showed the first evidence of Zn biomagnification. They did conclude that further laboratory and field studies were necessary to understand the precise mechanisms of metal trophic transfer [77]. Fisher and Reinfelder [78] discovered that concentrations of Cd increased from phytoplankton to zooplankton but decreased in fish, indicating lower trophic level biomagnification. Larison et al. [79] explain how herbivores such as beaver or muskrat, whose preferred food is any type of willow (*Salix spp.),* whether that be ground-hugging, alpine, or from montane marshes, are in danger of ingesting too much Cd. Willows act as biological pumps where cadmium is biomagnified, concentrating the metal by two orders of magnitude above background levels. Chen et al. [77] discovered that in freshwater environments, both diet and exposure to aquatic concentrations are significant sources of metals for zooplankton, but for fish, the sources vary among metals. These studies shed light on the complexities associated biomagnification of metals and that further research is required. According to Scheuhammer et al. [37], mink and otter, both top predators in an aquatic food chain, are at greater risk for elevated dietary MeHg exposure, accumulation, and toxicity. Greater exposure to metals is a problem for piscivorous wildlife in inland freshwater habitats compared to marine or estuarine habitats [37].

6.3 Body condition

Body condition can be defined as the energy capital accumulated in an animal's body from feeding, which is an indicator of health and quality of life [80]. Study results may also indicate a relationship between metals and individual body condition. If metal exposure affects body condition, then individuals with higher metal levels in their tissues may have lower body condition scores, which in many species may be associated with lower reproductive success [81, 82]. For the calculation of a CI (condition index) to be successful, it should take into consideration the fundamentals of growth, which are time and mass [80]. Calculating CI can help in understanding metal effects on an individual's health. Heavy metals are known to cause many issues, which lead to a degradation in overall well-being and a loss of weight and body fat [21, 83, 84]. There are many methods for calculating CI, including the Soxhlet procedure for understanding body fat. This procedure involves grinding or mincing the carcass and drying the samples to determine a lipid index [85]. The following formula was used to calculate body condition in cottontail rabbits (*Sylvilagus floridanus*) in Illinois:

CI = (W-16) L

CI = Condition Index

W = Weight in grams

L = Length in decimeters

A body condition formula developed for aquatic mammals such as the otter (*Lutra lutra*), was developed by Kruuk and Conroy [53], which gives a low value for poor condition and a high value for good condition:

Body Condition = W(aL^n)

W = Mean body weight (kg)

a = 5.02

L = contour length (m)

n = 2.33

In cervids, antler beam diameter can be used to evaluate body condition [86, 87]. First described by Cheatum [88, 89], bone marrow is also a widely used index for nutritional status in animals. Considered to be the last fat source depleted in a poorly nourished animal, low bone marrow fat indicates poor nutrition over an extended period of time [85]. However, according to Tête et al. [83], caution should be observed when using indices to make conclusions on exposure to toxic contaminants as they are highly influenced by several biological and environmental parameters.

6.4 Animal tissues for biomonitoring

There are methods available for metal evaluation in wildlife. Tissue harvest as done by Parker and Hamr [90] is an excellent way to study metal accumulation in wildlife, as it gives a thorough understanding of accumulation in the body. Tissues such as muscle, heart, diaphragm, liver, kidney, and adipose are common tissues used for metal analysis [84, 91–93]. Additionally, Hunter and Johnson [38] have demonstrated that brain tissue can be used as well. However, tissue sample methods are invasive and requires an ethical source of carcasses. When this is not an option, studies such as Lepage and Parker [76], Hunter and Johnson [38] and Curi et al. [94] using hair samples provided adequate data as well. There is some criticism of this method as it was found to be too many temporal variations associated with different seasonal pelage. For avian species, eggs, feathers, and urine have been used with success for metal analysis [74, 95]. Different metals target different organs, which can impact overall body condition. Many studies show that metals tend to accumulate more in soft tissues such as liver and kidneys [30, 96]. In wild aquatic mammals, higher levels are assumed to be found in the kidney, the primary accumulator organ [50] and in the liver, an organ crucial for ecotoxicological studies [30]. There are limitations to consider when sourcing animals from the wild. Due to yearly trapping and hunting pressure, animals from popular trapping and hunting grounds are usually young animals (the most common age class), causing a weighted age effect in sample sizes. Additionally, when searching for contaminants that may impair motor coordination such as Hg, only animals active and healthy enough to look for bait set by trapper would have been captured skewing data results [97]. When analyzing the impact of metals on aquatic environments, most studies focus on fish as their indicator species due to their sensitive nature [98]. However, aquatic mammals from fresh water systems seem to be researched less, perhaps due to the ease of catching fish and difficulty of trapping and sampling mammals.

6.5 Soil and geology

Metals occur in small amounts naturally and are dependent to the geology of the area. They can enter aquatic systems through ore-bearing rocks, wind-blown sediments, volcanoes, forest fires, and metal-accumulating vegetation [75, 99]. Not all studies consider the geology of the area when evaluating metal presence in wildlife. This can lead to misunderstandings in what natural levels in tissues should be in a specific region. Metals are always present at natural non-anthropogenic levels. Their introduction into ecosystem can be from processes as simple as the weathering of parent rocks and pedogenesis, inputting metals into soils and up trophic levels [100]. Wren et al. [101] for example, found different tissue metal concentrations in wild animals such as mink and river otter that can be attribute to natural geological and regional differences. Holmgren et al. [102] studied the natural differences of metals across the United States and, using GIS, found observable patterns of natural metal levels associated with different landscape types, from valleys to coastal ranges. For the province of Ontario, natural chemical and metal levels for soils and vegetation are publicly accessible for reference through the Ministry of Environment and Energy [103]. To make a thorough assessment of metal impact on wildlife, a distinction needs to be made between the effects of metals related to anthropogenic influences and those related to soil and geological systems in the area being assessed [27].

7. General conclusions and available methods

Metals originating from the environment are re-introduced, often in bioavailable forms from anthropogenic activities [104–106]. Their movement throughout our natural systems is not always visible, making them difficult to study and understand. Although the effects of individual metals has been studied more, the amalgamated effects of multiple metals on individuals are not well understood. There are very few studies on the effects of many metals from multiple trophic levels. Understanding this level of ecotoxicology can be useful in understanding the in-situ effects that urban development and industries such as mining and forestry have on the local environment. The primary value of this paper is that the various methods that have been used in the original manuscripts are available to researchers as well as the conservation measures and concerns that society and governments have identified.

Acknowledgements

We thank the Goodman School of Mines and Laurentian University for financing this research, the Timmins Fur Trappers Association, the Mattagami Metis Association, and the Friends of the Porcupine River for logistics, and Alexandra Binns, Cassidy Sullivan, and Fatimatzahra Muhammad for technical support. This manuscript is based on Carlie Stransky Ogilivie's Masters of Science thesis and has not been published previously.

DOI: http://dx.doi.org/10.5772/intechopen.1007359

Author details

Carlie Stransky Ogilvie[1], Graeme A. Spiers[2] and Frank F. Mallory[1]*

1 School of Natural Sciences, Laurentian University, Sudbury, Ontario, Canada

2 Harquail School of Earth Science, Laurentian University, Sudbury, Ontario, Canada

*Address all correspondence to: fmallory@laurentian.ca

References

[1] Jundt T. Dueling visions for the postwar world: The UN and UNESCO 1949 conferences on resources and nature, and the origins of environmentalism. Journal of American History. 2014;**101**(1):44-70. DOI: 10.1093/jahist/jau246

[2] Meyers-Schone L, Shugart LR, Beauchamp JJ, Walton BT. Comparison of two freshwater turtle species as monitors of radionuclide and chemical contamination: DNA damage and residue analysis. Environmental Toxicology and Chemistry. 1993;**12**:1487-1496

[3] Walker K, Walsh L. "No one yet knows what the ultimate consequences may be": How Rachel Carson transformed scientific uncertainty into a site for public participation in silent spring. Journal of Business and Technical Communication. 2012:26. DOI: 10.1177/1050651911421122

[4] Carson R. Silent Spring. Boston: Houghton Mifflin Publishers; 1962

[5] Rattner BA, Scheuhammer AM, Elliott JE. History of wildlife toxicology and the interpretation of contaminant concentrations in tissues. In: Environmental Contaminants in Biota: Interpreting Tissue Concentrations. CRC Press; 2011. pp. 9-44

[6] Heath RG, Spann JW, Hill EF, Kreitzer JF. Comparative dietary toxicities of pesticides to birds. U.S. Fish and Wildlife Service. 1972;**57**:152

[7] Schafer EWJ, Bowles WAJ, Hurlbut J. The acute oral toxicity, repellency, and hazard potential of 998 chemicals to one or more species of wild and domestic birds. Archives of Environmental Contamination and Toxicology. 1983;**12**:355-382

[8] Caro TM, O'Doherty G. On the use of surrogate species in conservation biology. Conservation Biology. 1999;**13**(4):805-814. DOI: 10.1046/j.1523-1739.1999.98338.x

[9] Environment Canada: Acts, Regulations and Agreements. 2016. Available from: http://www.ec.gc.ca/default.asp?lang=En&n=48d356c1-1

[10] Garson J, Aggarwal A, Sarkar S. Birds as surrogates for biodiversity: An analysis of a data set from southern Québec. Journal of Biosciences. 2002;**27** (4 Suppl. 2):347-360

[11] Burgas D, Byholm P, Parkkima T. Raptors as surrogates of biodiversity along a landscape gradient. Journal of Applied Ecology. 2014;**51**(3):786-794. DOI: 10.1111/1365-2664.12229

[12] Angelovicova L, Lodenius M, Tulisalo E, Fazekasova D. Effect of heavy metals on soil enzyme activity at different field conditions in middle spis mining area (Slovakia). Bulletin of Environmental Contamination and Toxicology. 2014;**93**:670-675. DOI: 10.1007/s00128-014-1397-0

[13] Shin MY, Cho YE, Park C, Sohn HY, Lim JH, Kwun IS. The contents of heavy metals (Cd, Cr, As, Pb, Ni, and Sn) in the selected commercial yam powder products in South Korea. Preventive Nutrition and Food Science. 2013;**18**(4):249-255. DOI: 10.3746/pnf.2013.18.4.249

[14] Beyer WN, Meador JP. Environmental Contaminants in Biota: Interpreting Tissue Concentrations. CRC Press; 2011. p. 601

[15] Calvert JH. Pheasant Poisoning by Swallowing Shot. Canadian Soil

Quality Guidelines for the Protection of Environmental and Human Health—Mercury (Inorganic). Canadian Environmental Quality Guidelines; 1999/1876. pp. 1-8

[16] Grinelle GB. Lead poisoning. Forest and Stream;**42**(6):117-118

[17] Haas JE, Hunter JE, Hamlin R, Taylor RJ, Brattin BL. Metal concentrations in tissues of mountain beaver (*Aplodontia rufa*). Northwestern Naturalist. 2009, 1894;**90**(3):194-206

[18] Stickel LF. Pesticide residues in birds and mammals. In: Edwards CA, editor. Environmental Pollution by Pesticides. London: Plenum Press; 1973. pp. 466-571

[19] Wetmore A. Lead Poisoning in Waterfowl. U.S. Department of Agriculture; 1919. pp. 793-712

[20] Magath TB. Lead poisoning in wild ducks. Proceedings of the Staff Management of the Mayo Clinic. 1931;**6**:749-752

[21] Jakimska A, Konieczka P, Skóra K, Namieśnik J. Bioaccumulation of metals in tissues of marine animals, part I: The role and impact of heavy metals on organisms. Polish Journal of Environmental Studies. 2011;**31**:22-39

[22] Jewett SC, Naidu AS. Assessment of heavy metals in red king crabs following offshore placer gold mining. Marine Pollution Bulletin. 2000;**40**(6):478-490. DOI: 10.1016/S0025-326X(00)00037-0

[23] Saunders JR, Knopper LD, Yagminas A, Koch I, Reimer KJ. Use of bio-markers to show sub-cellular effects in meadow voles (*Microtus pennsylvanicus*) living on an abandoned gold mine site. Science of the Total Environment. 2009;**407**(21):5548-5554. DOI: 10.1016/j.scitotenv.2009.07.026

[24] Wei-Chun M. Lead in mammals. In: Beyer WN, Meador JP, editors. Environmental Contaminants in Biota: Interpreting Tissue Concentrations. 2nd ed. CRC Press; 2011. p. 595

[25] Donato DB, Nichols O, Possingham H, Moore M, Ricci PF, Noller BN. A Critical Review of the Effects of Gold Cyanide-Bearing Tailings Solutions on Wildlife; 2007. pp. 42-56. doi:10.1016/j.envint.2007.04.007

[26] Linnik PM, Zubenko IB. Role of bottom sediments in the secondary pollution of aquatic environments by heavy-metal compounds. Lakes and Reservoirs: Research and Management. 2000;**5**(1):11-21. DOI: 10.1046/j.1440-1770.2000.00094.x

[27] Vries W, Römkens PFAM, Schütze G. Critical soil concentrations of cadmium, lead, and mercury in view of health effects on humans and animals. Reviews of Environmental Contamination and Toxicology. 2007a;**191**:91-130. DOI: 10.1007/978-0-387-69163-3_4

[28] Indian and Northern Affairs. Northwest Territories Contaminants Fact Sheet: Beaver and Muskrat; 2004. pp. 356-381

[29] Griffiths SR, Smith GB, Donato DB, Gillespie CG. Factors influencing the risk of wildlife cyanide poisoning on a tailings storage facility in the eastern goldfields of Western Australia. Ecotoxicology and Environmental Safety. 2009;**72**(5):1579-1586. DOI: 10.1016/j.ecoenv.2009.02.010

[30] Petkovšek S, Kopušar N, Kryštufek B. Small mammals as bio-monitors of metal pollution: A case study in Slovenia. Environmental Monitoring and Assessment. 2014;**186**:4261-4274

[31] Håkanson L, Nilsson A, Andersson T. Mercury in fish in Swedish

lakes. Environmental Pollution. 1988;**49**(2):145-162. DOI: 10.1016/0269-7491(88)90246-1

[32] DiFranco J, Bacon L, Mower B, Courtemanch D. Fish tissue contamination in Maine lakes. State of Maine Department of Environmental Protection; 1995. pp. 37-69

[33] Yeardley RB, Lazorchak JM, Paulsen SG. Elemental fish tissue contamination in northeastern U.S. lakes: Evaluation of an approach to regional assessment. Environmental Toxicology and Chemistry. 1998;**17**(9):1875-1884

[34] Miller P, Munkittrick KR, Dixon DG. Relationship between concentrations of copper and zinc in water, sediment, benthic invertebrates, and tissues of white sucker (*Catostomus commersoni*) at meta -contaminated sites. Canadian Journal of Fisheries and Aquatic Sciences. 1992;**49**:978-984

[35] Eisler R. A review of arsenic hazards to plants and animals with emphasis on fishery and wildlife resources. In: Ecotoxicology of Amphibians and Reptiles. 1994. pp. 185-259. Available from: https://pubs.er.usgs.gov/publication/5210532

[36] Suedel BC, Boraczek JA, Peddicord RK, Clifford PA, Dillon TM. In: Smith GW, editor. Trophic Transfer and Biomagnification Potential of Contaminants in Aquatic Ecosystems. 1994. pp. 3-34

[37] Scheuhammer AM, Meyer MW, Sandheinrich MB, Murray MW. Effects of methylmercury on the health of wild birds, mammals and fish. Royal Swedish Academy of Sciences. 2007;**36**(1):12-18. Available from: http://www.jstor.org/stable/4315780

[38] Hunter BA, Johnson MS. Food chain relationships of copper and cadmium in contaminated grassland ecosystems. Oikos. 1982;**38**(1):108-117. Available from: http://www.jstor.org/stable/3544572

[39] Khayatzadeh J, Abbasi E. The Effects of Heavy Metals on Aquatic Animals. The 1st International Applied Geological Congress, Department of Geology. Vol. 1. Iran: Islamic Azad University–Mashad Branch; 2010. pp. 26-28

[40] Tchounwou PB, Yedjou CG, Patlolla AK, Sutton DJ. Heavy Metals Toxicity and the Environment; 1967. pp. 6-53. 10.1007/978-3-7643-8340-4_6

[41] Mcgraw KJ. Melanins metals and mates. Oikos. 2003;**102**(2):402-406. Available from: http://www.jstor.org/stable/354804331

[42] Gál J, Hursthouse A, Tatner P, Stewart F, Welton R. Cobalt and secondary poisoning in the terrestrial food chain: Data review and research gaps to support risk assessment. Environment International. 2008;**34**(6):821-838. DOI: 10.1016/j.envint.2007.10.006

[43] ATSDR. Public Health Statement: Cobalt. Department of Health and Human Services; 2004, 2004. pp. 500-553

[44] De Boeck M, Kirsch-Volders M, Lison D. Cobalt and antimony: Genotoxicity and carcinogenicity. Mutation Research—Fundamental and Molecular Mechanisms of Mutagenesis. 2003;**533**(1-2):135-152. DOI: 10.1016/j.mrfmmm.2003.07.012

[45] Robison SH, Cantoni O, Costa M. Strand breakage and decreased molecular weight of DNA induced by specific metal compounds. Carcinogenesis. 1982;**3**(6):657-662. DOI: 10.1093/carcin/3.6.657

[46] Mudhoo A, Garg VK, Wang S. Removal of heavy metals by biosorption. Environmental Chemistry Letters. 2012;**4**:20-31. DOI: 10.1007/s10311-011-0342-2

[47] Goyer RA, Clarkson TW. Toxic effects of metals. In: Casarett and Doull's Toxicology, the Basic Science of Poison. 2001. pp. 811-867

[48] Hoffmann SR, Blunck SA, Petersen KN, Jones EM, Koval JC, Misek R, et al. Cadmium, copper, iron, and zinc concentrations in kidneys of grey wolves, *Canis lupus,* from Alaska, Idaho, Montana (USA) and the Northwest Territories (Canada). Bulletin of Environmental Contamination and Toxicology. 2010;**85**(5):481-485. DOI: 10.1007/s00128-010-0126-6

[49] Hayat S, Javed M, Razzao S. Growth performance of metal stressed. Pakistan Veterinary Journal. 2007;**27**(1):8-12

[50] Parker GH. Tissue metal levels in muskrat (*Ondatra zibethica*) collected near the Sudbury (Ontario) ore-smelters; prospects for biomonitoring marsh pollution. Environmental Pollution. 2004;**129**(1):23-30. DOI: 10.1016/j.envpol.2003.10.003

[51] DeForest DK, Schlekat CE, Brix KV, Fairbrother A. Secondary poisoning risk assessment of terrestrial birds and mammals exposed to nickel. Integrated Environmental Assessment and Management. 2012;**8**(1):107-119. DOI: 10.1002/ieam.248

[52] Ralston NVC, Raymond LJ. Dietary selenium's protective effects against methylmercury toxicity. Toxicology. 2010;**278**(1):112-123. DOI: 10.1016/j.tox.2010.06.004

[53] Kruuk H, Conroy JWH. Mortality of otters (*Lutra lutra*) in shetland. Journal of Applied Ecology. 1991;**28**(1):83-94. Available from: http://www.jstor.org/stable/2404115

[54] Ikemoto T, Kunito T, Anan Y, Tanaka H. Association of heavy metals with metallothionein and other proteins in hepatic cytosol of marine mammals and seabirds. Environmental Toxicology and Chemistry. 2004;**23**(8):2008-2016

[55] Khan MAK, Wang F. Mercury-selenium compounds and their toxicologiucal significance: Toward a molecular understanding of the mercury-selenium antagonism. Environmental Toxicology and Chemistry. 2009;**28**(8):1567. DOI: 10.1897/08-375.1

[56] Haines KJR, Evans RD, O'Brien M, Evans HE. Accumulation of mercury and selenium in the brain of river otters (*Lontra canadensis*) and wild mink (*Mustela vison*) from Nova Scotia, Canada. Science of the Total Environment. 2010;**408**(3):537-542. DOI: 10.1016/j.scitotenv.2009.09.055

[57] Whanger PD. Metabolic interactions of selenium with cadmium, mercury, and silver. Advances in Nutritional Research. 1985;**7**:221-250

[58] Drouhot S, Raoul F, Crini N, Tougard C, Prudent AS, Druart C, et al. Responses of wild small mammals to arsenic pollution at a partially remediated mining site in southern France. Science of the Total Environment. 2014:470-471. DOI: 10.1016/j.scitotenv.2013.10.053

[59] Espinosa-Reyes G, Torres-Dosal A, Ilizaliturri C, Gonzalez-Mille D, Diaz-Barriga F, Mejia-Saavedra J. Wild rodents (*Dipodomys merriami*) used as bio-monitors in contaminated mining sites. Journal of Environmental Science and Health Part A, Toxic/Hazardous Substances and Environmental Engineering. 2010;**45**(1):82-89. DOI: 10.1080/10934520903388988

[60] Driscoll CT, Mason RP, Chan HM, Jacob DJ, Pirrone N. Mercury as a global pollutant: Sources, pathways, and effects. Environmental Science and Technology. 2013;**47**(10):4967-4983. DOI: 10.1021/es305071v

[61] Klenavic K, Champoux L, Mike O, Daoust PY, Evans RD, Evans HE. Mercury concentrations in wild mink (*Mustela vison*) and river otters (*Lontra canadensis*) collected from eastern and Atlantic Canada: Relationship to age and parasitism. Environmental Pollution. 2008;**156**(2):359-366. DOI: 10.1016/j.envpol.2008.02.003

[62] Wiener JG, Krabbenhoft DP, Heinz GH, Scheuhammer AM. Ecotoxicology of mercury. In: Hoffman DJ, Rattner BA, Burton GAJ, Cairns J, editors. Handbook of Ecotoxicology. Boca Raton FL: CRC Press; 2003. pp. 409-463

[63] Oliveira Ribeiro CA, Rouleau C, Pelletier E, Audet C, Tjalve H. Distribution kenetics of dietary methylmercury in the arctic char (*Salvelinus alpinus*). Environmental Science and Technology. 1999;**33**:902-907

[64] Sandheinrich MB, Wiener JG. Methylmercury in freshwater fish: Recent advances in assessing toxicity of environmentally relevant exposures. In: Beyer WN, Meador JP, editors. Environmental Contaminants in Biota: Interpreting Tissue Concentrations. 2nd ed. Boca Raton, FL: CRC Press; 2011. pp. 169-190

[65] Engstrom D, Feng X, Fitzgerald W, Pirrone N, Prestbo E, Seigneur C. The Madison declaration on mercury pollution. AMBIO: A Journal of the Human Environment. 1 Feb 2007;**36**(1):62-66. DOI: 10.1579/0044-7447(2007)36[62:TMDOMP]2.0.CO;2

[66] UNEP. Global Mercury Assessment 2013: Sources, Emissions, Releases and Environmental Transport. UNEP Chemical Branch. Geneva, Switzerland: Narayana Press; 2013. pp. 200-206. DOI: 10.1016/S0300-483X(03)00203-8

[67] Keita-Ouane F. UNEP Chemicals' Work: Breaking the Barriers to Information Access. Toxicology. Geneva, Switzerland: Narayana Press; 2003. pp. 20-45. DOI: 10.1016/S0300-483X(03)00203-8

[68] Cooke JA. Cadmium in small mammals. In: Beyer WN, Meador JP, editors. Environmental Contaminants in Biota: Interpreting Tissue Concentrations. Boca Raton, FL: CRC Press; 2011. pp. 627-639

[69] Hopkins WA, Hopkins LB, Unrine JM, Snodgrass J, Elliot JD. Mercury concentrations in tissues of osprey from the Carolinas, USA. Journal of Wildlife Management. 2007;**71**(6):1819-1829. DOI: 10.2193/2006-016

[70] Talmage SS, Walton BT. Small mammals as monitors of environmental contaminants. Environmental Pollution. 1991;**113**(3):347-355. DOI: 10.1016/S0269-7491(00)00183-4

[71] Sánchez-Chardi A, Ribeiro CAO, Nadal J. Metals in liver and kidneys and the effects of chronic exposure to pyrite mine pollution in the shrew (*Crocidura russula*) inhabiting the protected wetland of Doñana. Chemosphere. 2009;**9**:23-46. DOI: 10.1016/j.chemosphere.2009.03.036

[72] Hallén IP, Jorhem L, Oskarsson A. Placental and lactational transfer of lead in rats: A study on the lactational process and effects on offspring. Archives of Toxicology. 1995;**69**(9):596-602. Available from: http://www.ncbi.nlm.nih.gov/pubmed/8660136

[73] Milton A, Cooke JA, Johnson MS. Accumulation of lead, zinc, and cadmium in a wild population of *Clethrionomys glareolus* from an abandoned lead mine. Archives of Environmental Contamination and Toxicology. 2003;**44**(3):405-411. DOI: 10.1007/s00244-002-2014-5

[74] Rose G, Parker GH. Effects of smelter emissions on metal levels in the plumage of ruffed grouse near Sudbury, Ontario, Canada. Canadian Journal of Zoology. 1982;**60**:2659-2667. DOI: 10.1139/z82-342

[75] Ccme. Canadian soil quality guidelines for the protection of environmental and human health—Cadmium. Canadian Environmental Quality Guidelines. 1999;**9**:34-67

[76] Lepage P, Parker GH. Copper, nickel, and iron levels in pelage of red squirrels living near the ore smelters at Sudbury, Ontario, Canada. Canadian Journal of Zoology. 1988;**66**(7):1631-1637. DOI: 10.1139/z88-238

[77] Chen CY, Stemberger RS, Kamman NC, Mayes BM, Folt CL. Patterns of Hg bioaccumulation and transfer in aquatic food webs across multi-lake studies in the northeast US. Ecotoxicology. 2005;**14**:135-147. DOI: 10.1007/s10646-004-6265-y

[78] Fisher NS, Reinfelder JR. The trophic transfer of metals in marine systems. In: Tessier DR, Turner A, editors. Metal Speciation and Bioavailability in Aquatic Systems. Chichester: John Wiley and Sons; 1995. pp. 363-406

[79] Larison JR, Likens GE, Fitzpatrick JW, Crock JG. Cadmium toxicity among wildlife in the Colorado rocky mountains. Nature. 2000;**406**(6792):181-183. DOI: 10.1038/35018068

[80] Peig J, Green AJ. New perspectives for estimating body condition from mass/length data: The scaled mass index as an alternative method. Oikos. 2009;**118**(12):1883-1891. DOI: 10.1111/j.1600-0706.2009.17643.x

[81] Hall MN. Parameters associated with cyclic populations of arctic fox (*Alopex lagopus*) near Eskimo point, Northwest Territories: Age, condition, seasonal and multiannual influences [thesis]. Laurentian University M.Sc; 1989

[82] Mallory FF, Hall MN. Condition indices of cyclic populations of arctic fox near Eskimo point, Northwest Territories: Relationship to morphology, age, season and year. In: Proceedings of the 2nd International Conference of the North, Tyumen, Russia. 1993

[83] Tête N, Fritsch C, Afonso E, Coeurdassier M, Lambert JC, Giraudoux P, et al. Can body condition and somatic indices be used to evaluate metal-induced stress in wild small mammals? PLoS One. 2013;**8**(6):37-45. DOI: 10.1371/journal.pone.0066399

[84] Giżejewska A, Spodniewska A, Barski D, Fattebert J. Beavers indicate metal pollution away from industrial centers in northeastern Poland. Environmental Science and Pollution Research. 2015;**22**:3969-3975. DOI: 10.1007/s11356-014-3769-8

[85] Harder DJ, Kirkpatrick RL. Physiological Methods in Wildlife Research. Boca Raton, FL: CRC Press; 1996, 1996. p. 695

[86] Severinghaus CW, Maguire HF, Cookingham RA, Tanck JE. Variations by age class in the antler beam diameters of white tailed deer related to range condition. Transactions of the North American Wildlife Conference. 1950;**15**:551-570

[87] Riney T. Evaluating condition of the freerange red deer (*Cervus elaphus*), with special reference to New Zealand. New Zealand Journal of Science Technolology Section B. 1955;**36**:429-463

[88] Cheatum EL. Bone marrow as an index of malnutrition in deer. N.Y. State Conservation. 1949;**3**:19-22

[89] Cheatum EL. Bone marrow as an index of malnutrition in African ungulates. South African Journal of Wildlife Research. 1977;**7**(2):61-66

[90] Parker GH, Hamr J. Metal levels in body tissues, forage and fecal pellets of elk (*Cervus elaphus*) living near the ore smelters at Sudbury, Ontario. Environmental Pollution. 2001;**113**:347-355. DOI: 10.1016/S0269-7491(00)00183-4

[91] Bennett PM, Jepson PD, Law RJ, Jones BR, Kuiken T, Baker JR, et al. Exposure to heavy metals and infectious disease mortality in harbour porpoises from England and Wales. Environmental Pollution. 2001;**112**:33-40. DOI: 10.1016/S0269-7491(00)00105-6

[92] Alleva E, Francia N, Pandolfi M, De Marinis AM, Chiarotti F, Santucci D. Organochlorine and heavy-metal contaminants in wild mammals and birds of Urbino/Pesaro Province, Italy: An analytic overview for potential bioindicators. Archives of Environmental Contamination and Toxicology. 2006:396-478. DOI: 10.1007/s00244-005-0218-1

[93] Rosa C, Blake JE, Bratton GR, Dehn LA, Gray MJ, O'Hara TM. Heavy metal and mineral concentrations and their relationship to histopathological findings in the bowhead whale (*Balaena mysticetus*). Science of the Total Environment. 2008;**399**(1-3):165-178. DOI: 10.1016/j.scitotenv.2008.01.062

[94] Curi NHDA, Brait CHH, Filho NRA, Talamoni SA. Heavy metals in hair of wild canids from the Brazilian Cerrado. Biological Trace Element Research. 2012;**147**:97-102. DOI: 10.1007/s12011-011-9303-7

[95] Clapp JB, Bevan RM, Singleton I. Avian urine: Its potential as a non-invasive biomonitor of environmental metal exposure in birds. Water, Air, and Soil Pollution. 2012;**6**:39-52. DOI: 10.1007/s11270-012-1161-1

[96] Amuno S, Niyogi S, Amuno M, Attitaq J. Heavy metal bioaccumulation and histopathological alterations in wild Arctic hares (*Lepus arcticus*) inhabiting a former lead/zinc mine in the Canadian high Arctic. Science of the Total Environment. 2016;**7**:59-72. DOI: 10.1016/j.scitotenv.2016.03.007

[97] Mierle G, Addison EM, MacDonald KS, Joachim DG. Mercury levels in tissues of otters from Ontario, Canada: Variations with age, sex, and location. Environmental Toxicology. 2000;**19**:3044-3051

[98] Tayefi-Nasrabadi H, Rahmani R. Partial purification and characterization of rhodanese from rainbow trout (*Oncorhynchus mykiss*) liver. The Scientific World Journal. 2012:648085. DOI: 10.1100/2012/648085

[99] Fernandez-Leborans G, Olalla Herrero Y. Toxicity and bioaccumulation of lead and cadmium in marine protozoan communities. Ecotoxicology and Environmental Safety. 2000;**47**(3):266-276. DOI: 10.1006/eesa.2000.1944

[100] Facchinelli A, Sacchi E, Mallen L. Multivariate statistical and GIS-based approach to identify heavy metal sources in soils. Environmental Pollution. 2001;**114**:313-324. DOI: 10.1016/S0269-7491(00)00243-8

[101] Wren CD, Fischer KL, Stokes PM. Levels of lead, cadmium and other elements in mink and otter from Ontario, Canada. Environmental Pollution. 1988;**52**:193-202. DOI: 10.1016/0269-7491(88)90003-6

[102] Holmgren GGS, Meyer MW, Chaney RL, Daniels RB. Cadmium, lead, zinc, copper, and nickel in agricultural soils of the United States of America. Journal of Environment Quality. 1993;**22**:335. DOI: 10.2134/jeq1993.00472425002200020015x

[103] MOEE. Ontario Typical Range of Chemical Parameters in Soil, Vegetation, Moss Bags and Snow. 1st ed. Boca Raton, FL: CRC Press; 1993, 1993. pp. 595-603

[104] Parker GH, Van Der Kraak G. Use of metal-rich slag as a source of grit and its effects on pigeon health and fitness. Journal of Environmental Science and Pollution Research. 2018;**4**:256-267

[105] Song SY, Lee S, Lee H, Nam MJ, Choi J. Integrated geochemical and geophysical monitoring for the assessment of natural pollutant mitigation in heavy metal-contaminated areas of South Korea, Deagu. Journal of Environmental Science and Pollution Research. 2024;**31**:45650-45666. DOI: 10.1007/s11356-024-34008-z

[106] Scarfone A, Cammerata A, Romano E, Vinciguerra V, Marabottini R, Gallucci F, et al. Effects of ground transport on the presence of heavy metals in selected honey bee products. Environmental Science and Pollution Research. Jun 2024;**31**(30):1-12. DOI: 10.1007/s11356-024-33982-8